AF836836

ENGINEERING CHEMISTRY

Asmae Bouziani

Toronto Academic Press

ENGINEERING CHEMISTRY

Asmae Bouziani

Toronto Academic Press
4164 Lakeshore Road
Burlington ON L7L 1A4
Canada
www.tap-books.com

Email: orders@arclereducation.com

ISBN: 978-1-77956-305-7

ABOUT THE AUTHOR

Asmae Bouziani, Ph.D., is an accomplished physical chemist with a specialization in hydrogen production, catalytic dehydrogenation, and photocatalysis. Dr. Bouziani's research is at the cutting edge of material science, particularly in developing advanced photocatalytic systems aimed at improving efficiency and sustainability in chemical processes. Her work has significantly contributed to the enhancement of photocatalytic technologies, making strides in both theoretical understanding and practical applications. Dr. Bouziani earned her doctorate in Physical Chemistry and has since accumulated extensive postdoctoral experience, marked by numerous publications and conference presentations. She is also a dedicated educator, committed to teaching and mentoring the next generation of scientists. Fluent in Arabic, English, and French, with proficiency in Turkish and Spanish, Dr. Bouziani is committed to fostering international collaboration. Her current work reflects a deep interest in applying her analytical chemistry expertise to alternative energy research, further broadening the scope of her scientific contributions.

Table of Contents

List of Figures

List of Tables

PREFACE

Engineering Chemistry is a multidisciplinary field at the intersection of chemistry and engineering, focusing on the practical applications of chemical principles in various engineering contexts. It encompasses the study of chemical processes, reactions, and materials crucial for designing, developing, and optimizing technological solutions across industries. From the molecular level to large-scale production systems, engineering chemistry plays a pivotal role in ensuring the efficiency, sustainability, and safety of processes in fields such as manufacturing, energy production, environmental engineering, and materials science. It integrates fundamental chemistry with engineering principles to innovate and implement solutions that address global challenges, ranging from clean energy technologies to advanced materials development.

This book, *Engineering Chemistry*, presents a thorough exploration across eight focused chapters, aiming to elucidate the fundamental principles and practical applications of chemistry within engineering contexts. **Chapter 1** introduces the *Fundamentals of Engineering Chemistry*, covering atomic structure, chemical bonding, and the basic laws governing chemical reactions essential for foundational understanding. **Chapter 2** dives into *Atomic Structure and Bonding*, detailing the structural arrangements of atoms and the bonding forces that dictate their interactions. **Chapters 3** and **4** explore *Inorganic Chemistry* and *Organic Chemistry*, respectively, outlining the properties and reactions of inorganic and organic compounds crucial for practical engineering applications.

In **Chapter 5**, *Chemical Kinetics and Thermodynamics* are discussed, focusing on reaction rates and energy transformations critical for designing and optimizing chemical processes. **Chapter 6** explores *Electrochemistry*, examining electrochemical reactions and their applications in fields such as energy storage and corrosion protection. **Chapter 7** addresses *Corrosion and Its Control*, emphasizing strategies to mitigate corrosion effects through material selection and protective coatings. Finally, **Chapter 8** delves into *Green Chemistry*, highlighting sustainable practices to minimize environmental impact in chemical processes.

This book is designed for students, researchers, and professionals seeking a comprehensive understanding of how chemistry underpins engineering practices and innovations. By integrating theoretical insights with practical applications throughout these chapters, the goal is to provide a valuable resource that supports innovation and sustainability in engineering chemistry.

—Author

CHAPTER

1 Fundamentals of Engineering Chemistry

LEARNING OBJECTIVES

After studying this chapter, you will be able to:

1. Understand emerging chemical technologies relevant to engineering applications.
2. Apply SI units and conversions accurately in chemical measurements.
3. Explain the significance of significant figures in measurement accuracy.
4. Define the mole concept and relate it to Avogadro's number.
5. Perform stoichiometric calculations and balance chemical equations.
6. Describe different types of solutions and their properties.
7. Analyze the pH scale and its importance in acid-base chemistry.

KEY TERMS FROM THIS CHAPTER

Acid-base titrations
Mole concept
SI units
Stoichiometry
Colligative properties
pH scale
Significant figures

INTRODUCTION

Chemistry plays a fundamental role in various engineering disciplines, providing the essential principles and tools required for the development and optimization of materials, processes, and technologies. Understanding the chemical properties and behaviors of different substances allows engineers to innovate and improve solutions across a wide range of fields, from materials science and environmental engineering to biomedical applications and energy production (Yang et al., 2018). In "materials science and engineering," chemistry is crucial for the development of new materials with specific properties, such as strength, flexibility, and resistance to corrosion. For example, the design of advanced composites, polymers, and nanomaterials relies heavily on chemical knowledge to tailor their molecular structures and interactions. In "environmental engineering," chemistry is vital for the treatment of pollutants, water purification, and the development of sustainable processes. Chemical reactions and interactions are the basis for removing contaminants from air, water, and soil, and for designing systems that minimize environmental impact (Dow, 1952).

In the "biomedical field," chemistry underpins the creation of pharmaceuticals, diagnostic tools, and medical devices. The synthesis of new drugs and the development of biomaterials for implants and prosthetics require a deep understanding of chemical principles. "Chemical engineering" itself is a discipline that merges principles of chemistry, physics, mathematics, and engineering to design and optimize processes for large-scale production of chemicals, fuels, and materials. The knowledge of chemical kinetics, thermodynamics, and reaction mechanisms is essential for designing efficient and safe industrial processes. The historical contributions of chemistry to engineering are significant and numerous. The "Industrial Revolution" was marked by the development of processes for the large-scale production of chemicals such as sulfuric acid and sodium carbonate, which were essential for various industrial applications. The "Haber-Bosch process," (Figure 1.1) developed in the early 20th century, revolutionized agriculture by enabling the synthesis of ammonia from nitrogen and hydrogen, leading to the production of synthetic fertilizers and dramatically increasing food production (Larson et al., 1994).

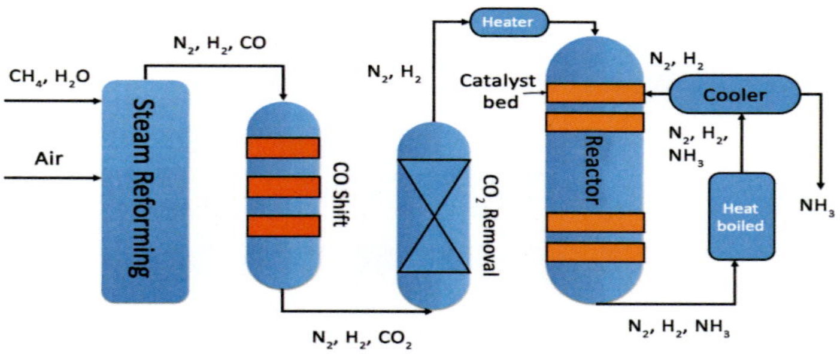

Figure 1.1. *Schematic of Haber-Bosch process.*

Source: CC BY, Creative Commons License.

The "development of synthetic polymers" in the 20th century, such as nylon, polyethylene, and Teflon, transformed industries ranging from textiles to aerospace. These materials, created through chemical reactions, offered new properties and functionalities that natural materials could not provide. The invention of "semiconductors" and the understanding of their chemical properties led to the development of the modern electronics industry, including transistors, integrated circuits, and microprocessors, which are the backbone of contemporary technology (Woods and Sawchuk, 1993).

1.1. EMERGING CHEMICAL TECHNOLOGIES IN ENGINEERING

The future of engineering is poised to benefit from several emerging chemical technologies. "Green chemistry" aims to design chemical products and processes that reduce or eliminate the use and generation of hazardous substances. This approach not only improves environmental sustainability but also enhances safety and efficiency in industrial processes. The development of "renewable energy sources," (Figure 1.2) such as solar cells, fuel cells, and biofuels, relies heavily on advancements in chemistry to create more efficient and cost-effective energy conversion and storage systems (Rubin et al., 2006).

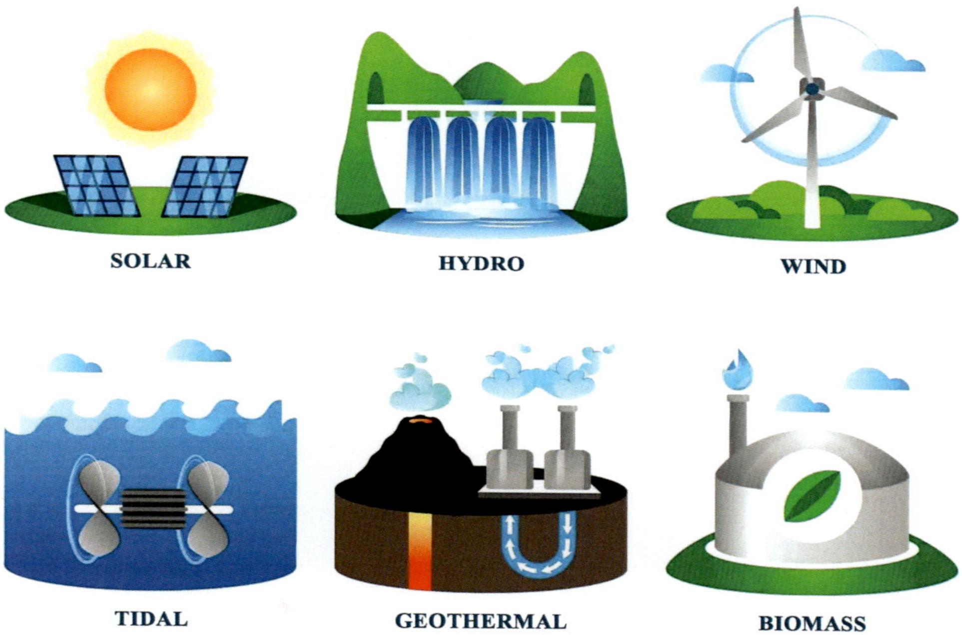

Figure 1.2. *Image of sources of renewable energy sources.*

Source: Investopedia, Creative Commons License.

Nanotechnology" is another frontier where chemistry plays a pivotal role. By manipulating matter at the atomic and molecular scale, engineers can develop materials with unprecedented properties, leading to innovations in electronics, medicine, and materials science. The field of "biotechnology" is also expanding, with chemical engineering principles being applied to genetic engineering, drug delivery systems, and the development of new biomaterials (Carpenter et al., 2017).

In conclusion, chemistry is integral to the advancement of engineering disciplines, providing the knowledge and tools necessary to develop new materials, processes, and technologies. The historical impact of chemistry on engineering has been profound, and future trends indicate that chemical innovations will continue to drive progress and address global challenges in sustainability, health, and technology (Li and Wang, 2020).

Remember

Remember to focus on the fundamental concepts discussed in this chapter, as they form the basis for understanding chemical processes in engineering.

1.2. BASIC CONCEPTS AND MEASUREMENT IN CHEMISTRY

Chemistry, as a quantitative science, relies heavily on precise measurement and the understanding of fundamental concepts. This section will cover the essential units of measurement, the significance of accuracy in data, the mole concept, and stoichiometry (Romine et al., 2016).

1.2.1. Units of Measurement: SI Units and Conversions

The International System of Units (SI) is the standard system of measurement used in chemistry and other scientific disciplines. It includes seven base units, such as the meter (m) for length, the kilogram (kg) for mass, the second (s) for time, and the mole (mol) for the amount of substance (Ford and Gilbert, 2013).

To convert between different units, a clear understanding of conversion factors is necessary. For instance, to convert meters to centimeters, one must know that 1 meter equals 100 centimeters:

$$1 \text{ m} = 100 \text{ cm}$$

Similarly, converting between mass units, such as grams and kilograms, involves the relation (Powsner, 1984).

$$1 \text{ kg} = 1000 \text{ g}$$

Conversion between temperature scales, like Celsius and Kelvin, is also common in chemistry (Flanagan, 1995).

$$T(K) = T(°C) + 273.15$$

1.2.2. Significant Figures and Accuracy in Measurements

Significant figures are the digits in a measurement that are known with certainty plus one digit that is estimated. They are crucial for indicating the precision of measurements and ensuring the accuracy of calculations. For example, in the measurement 23.45 grams, there are four significant figures (Gao and Lloyd, 2020).

Rules for significant figures:

1. Non-zero digits are always significant,
2. Any zeros between significant digits are significant,
3. Leading zeros are not significant, and
4. Trailing zeros in a decimal number are significant.

Accuracy refers to how close a measured value is to the true value, while precision indicates how closely repeated measurements are to each other. Accurate and precise measurements are critical for reliable data in chemical experiments (Bularzik, 2007).

Did you know?

Did you know that Avogadro's number (6.022 × 1023) represents the number of atoms or molecules in one mole of a substance? This constant is crucial for understanding the scale of chemical reactions.

1.2.3. The Mole Concept and Avogadro's Number

The mole is a fundamental unit in chemistry that represents a specific number of particles, typically atoms or molecules. Avogadro's number (6.022×10^{23}) defines the number of entities in one mole of a substance. This concept allows chemists to count particles by weighing them. For example, one mole of carbon-12 atoms has a mass of exactly 12 grams and contains 6.022×10^{23} atoms:

$$1 \text{ mol C} = 12 \text{ g C} = 6.022 \times 10^{23} \text{ atoms C}$$

The mole concept is essential for relating macroscopic measurements to the number of particles involved in chemical reactions (Bindel, 2021).

1.2.4. Stoichiometry: Balancing Chemical Equations and Calculations

Stoichiometry involves the calculation of reactants and products in chemical reactions. It is based on the conservation of mass, where the total mass of reactants equals the total mass of products. To balance a chemical equation, ensure that the number of atoms of each element is the same on both sides of the equation.

Tip

When performing stoichiometric calculations, always balance chemical equations first to ensure accurate measurements and predictions of reactant and product quantities.

For example, consider the combustion of methane:

$$CH_4 + 2O_2 \rightarrow CO_2 + 2H_{2O}$$

Here, one molecule of methane reacts with two molecules of oxygen to produce one molecule of carbon dioxide and two molecules of water. Stoichiometric calculations allow us to determine the amounts of reactants needed and products formed (Tóth and Sebestyén, 2009).

1.2.4.1. Balancing Chemical Equations

To balance chemical equations, follow these steps:

1. Write the unbalanced equation.
2. Count the number of atoms of each element on both sides of the equation.
3. Use coefficients to balance the atoms, starting with the most complex molecule.
4. Check that all atoms are balanced and adjust coefficients as necessary.

For instance, the combustion of ethane (C_2H_6):

$$C2H_6 \aleph O_2 \quad CO_2 \quad H_{2O}$$

Balancing (Bílek et al., 2018):

$$2C2H_6 + 7O_2 \rightarrow 4CO_2 + 6H_{2O}$$

1.2.4.2. Stoichiometric Calculations

Stoichiometry can be used to calculate the amount of reactants and products. For example, if 5 moles of ethane (C_2H_6) are burned, how many moles of CO_2 are produced?

From the balanced equation:

$$2C2H_6 + 7O_2 \rightarrow 4CO_2 + 6H_{2O}$$

The molar ratio of C_2H_6 to CO_2 is 2:4 or 1:2. Therefore, 5 moles of C_2H_6 will produce:

$$5 \text{ moles } C2H_6 \times \frac{4 \text{ moles } CO_2}{2 \text{ moles } C2H_6} = 10 \text{ moles } CO_2$$

In conclusion, understanding basic concepts and measurements in chemistry is crucial for accurate experimental results and effective communication in engineering. (Niaz and Montes, 2012).

1.2.5. States of Matter and Intermolecular Forces

Understanding the states of matter and the forces that hold them together is fundamental in chemistry. This section covers the properties of gases, liquids, and solids, as well as the intermolecular forces that influence their behavior (Widom, 1967).

1.2.5.1. Properties of Gases: Ideal Gas Law and Real Gases

Gases are characterized by their high compressibility and the ability to fill any container. The behavior of gases can be described by the Ideal Gas Law:

$$PV = nRT$$

where (P) is the pressure, (V) is the volume, (n) is the number of moles of gas, (R) is the universal gas constant (8.314 J/(mol·K)), and (T) is the temperature in Kelvin.

However, real gases deviate from ideal behavior under high pressure and low temperature. The Van der Waals equation accounts for these deviations:

$$\left(P+\frac{an^2}{V^2}\right)(V-nb)=nRT$$

where (a) and (b) are constants that correct for intermolecular forces and the finite volume of gas molecules, respectively (Gillespie, 1936).

1.2.5.2. Liquids: Vapor Pressure, Surface Tension, and Viscosity

Liquids have definite volume but take the shape of their container. Key properties of liquids include vapor pressure, surface tension, and viscosity:

1. **Vapor Pressure:** The pressure exerted by a vapor in equilibrium with its liquid at a given temperature. It increases with temperature as described by the Clausius-Clapeyron equation:

$$\ln\left(\frac{P_1}{P_2}\right)=\frac{\Delta H_{vap}}{R}\left(\frac{1}{T_2}-\frac{1}{T_1}\right)$$

where (P_1) and (P_2) are the vapor pressures at temperatures (T_1) and (T_2), and $(\Delta H_{ü}\ )$ is the enthalpy of vaporization.

2. **Surface Tension:** The energy required to increase the surface area of a liquid. It arises from the cohesive forces between molecules at the surface and is given by:

$$\gamma=\frac{F}{L}$$

where (γ) is the surface tension, (F) is the force, and (L) is the length over which the force acts.

3. **Viscosity:** A measure of a liquid's resistance to flow. It depends on the intermolecular forces and the structure of the liquid. The dynamic viscosity (η) is given by:

$$\eta=\frac{F}{A\cdot v}$$

where (F) is the force applied, (A) is the area, and (v) is the velocity gradient perpendicular to the flow direction (Haller et al., 2013).

1.2.5.3. Solids: Crystal Lattices, Amorphous Solids, and Allotropes

Solids have definite shape and volume. They can be classified into crystalline and amorphous solids:

1. **Crystal Lattices:** Crystalline solids have a regular, repeating arrangement of atoms, ions, or molecules. The basic repeating unit is called the unit cell. Common types of crystal lattices include face-centered cubic (FCC), body-centered cubic (BCC), and hexagonal close-packed (HCP).
2. **Amorphous Solids:** These lack a long-range ordered structure. Examples include glass and many plastics. They have irregular arrangements and do not exhibit the sharp melting points characteristic of crystalline solids.
3. **Allotropes:** Different structural forms of the same element in the

same physical state. For example, carbon has several allotropes including diamond, graphite, and graphene, each with distinct physical properties due to different bonding arrangements (Fahlman, 2023).

1.2.5.4. Intermolecular Forces

Intermolecular forces (Figure 1.3) are the forces of attraction between molecules. They are weaker than intramolecular forces (bonds within a molecule) but crucial for determining the physical properties of substances.

1. **Van der Waals Forces:** These include London dispersion forces, which arise from temporary dipoles in molecules, and are present in all molecular substances. The strength of these forces increases with the size and polarizability of the molecules.
2. **Dipole-Dipole Interactions:** Occur between polar molecules with permanent dipoles. The positive end of one dipole attracts the negative end of another. These interactions are stronger than Van der Waals forces and affect boiling points and solubilities.
3. **Hydrogen Bonding:** A special type of dipole-dipole interaction occurring when hydrogen is bonded to highly electronegative atoms like nitrogen, oxygen, or fluorine. Hydrogen bonds are significantly stronger than other dipole-dipole interactions and play a crucial role in the properties of water, biological molecules like DNA, and many other compounds (Tarhan et al., 2008).

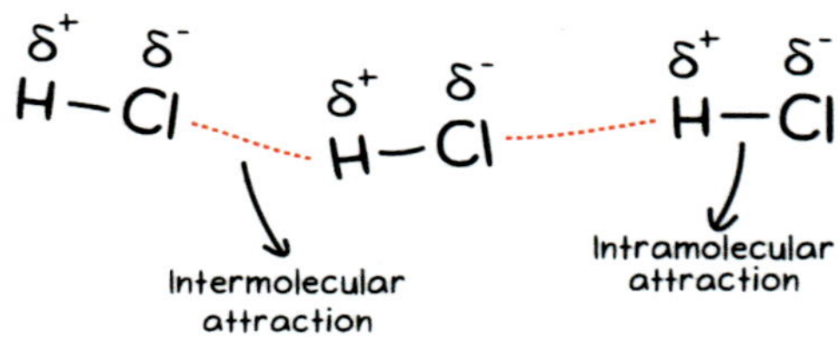

Figure 1.3. *Illustration of intermolecular forces.*

Source: Khan Academy, Creative Commons License.

1.3. SOLUTIONS AND THEIR PROPERTIES

A **solution** is a homogeneous mixture of two or more substances. In a solution, the substance present in the greatest amount is known as the **solvent**, and the substance or substances present in lesser amounts are known as the **solute(s)**. (Scatchard et al., 1957).

1.3.1. Types of Solutions

Solutions can be classified based on the states of their solute and solvent into solid, liquid, and gaseous solutions:

1. **Solid Solutions:** These occur when the solute and solvent are in the solid phase. An example is an alloy, such as brass (copper and zinc). The uniform distribution of atoms enhances the material's mechanical properties.
2. **Liquid Solutions:** The most common type, where the solute is dissolved in a liquid solvent. Examples include saltwater (solid solute in liquid solvent) and ethanol in water (liquid solute in liquid solvent). These solutions are essential in fields like pharmacology and chemical engineering.
3. **Gaseous Solutions:** These involve gases dissolving in other gases. Air is a primary example, consisting of nitrogen, oxygen, and other gases. Gas mixtures are vital in processes such as combustion and refrigeration (Zhang et al., 2010).

1.3.2. Concentration Units

Concentration quantifies the amount of solute in a given quantity of solvent or solution. Various units are used to express concentration:

1. Molarity (M): Defined as the number of moles of solute per liter of solution.

$$M = n_{\text{solute}} V \text{solution}$$

where $\left(n_{\text{solute}}\right)$ is the number of moles of solute, and $(V\text{solution}$ is the volume of the solution in liters. Molarity is temperature-dependent since volume changes with temperature.

2. Molality (m): Defined as the number of moles of solute per kilogram of solvent.

   $m = n_{\text{solute}} m\text{solvent}$

   where (n_{solute}) is the number of moles of solute, and $(m\text{solvent})$ is the mass of the solvent in kilograms. Molality is temperature-independent, making it useful for precise calculations in thermodynamics.

3. Normality (N): Measures the number of equivalents of solute per liter of solution.

   $N = n_{\text{equivalents}} V\text{solution}$

   where $(n_{\text{equivalents}})$ is the number of equivalents of solute. The concept of equivalents varies with the type of reaction, such as acid-base titrations or redox reactions (Korbag and Saleh, 2017).

1.3.3. Colligative Properties

Colligative properties depend on the number of solute particles in a solution, not their identity. They include boiling point elevation, freezing point depression, and osmotic pressure:

1. **Boiling Point Elevation:** The boiling point of a solution is higher than that of the pure solvent.

   $\Delta T_b = i \cdot K_b \cdot m$

   where (ΔT_b) is the boiling point elevation, (i) is the van 't Hoff factor (number of particles the solute dissociates into), (K_b) is the ebullioscopic constant, and (m) is the molality of the solution.

2. **Freezing Point Depression**: The freezing point of a solution is lower than that of the pure solvent.

   $\Delta T_f = i \cdot K_f \cdot m$

   where (ΔT_f) is the freezing point depression, (K_f) is the cryoscopic constant, and (m) is the molality of the solution.

3. **Osmotic Pressure:** The pressure required to stop the flow of solvent through a semipermeable membrane.

   $\Pi = i \cdot M \cdot R \cdot T$

   where (Đ) is the osmotic pressure, (M) is the molarity of the solution, (R) is the gas constant, and (T) is the temperature in Kelvin (Manning, 1969).

1.4. ACIDS, BASES, AND SALTS

Acids, bases, and salts play a pivotal role in various chemical processes and industrial applications. This section explores their definitions, theoretical frameworks, measurement techniques, practical applications, and their significance in engineering.

1.4.1. Arrhenius, Brønsted-Lowry, and Lewis Concepts

Understanding acids and bases requires a grasp of several fundamental theories:

1. Arrhenius Concept
   i. Acids: Substances that increase the concentration of hydrogen ions (H^+) in aqueous solutions.
   ii. Bases: Substances that increase the concentration of hydroxide ions (OH^-) in aqueous solutions (Andersen and Mack, 2017).

Example: $\left(HCl \rightarrow H^+ + Cl^-\right) \pm \left(NaOH \rightarrow Na^+ + OH^-\right)$.

2. **Brønsted-Lowry Concept** (Figure 1.4):
   i. Acids: Proton donors.
   ii. Bases: Proton acceptors.

Example: $\left(NH_3 + H_{2O} \rightleftharpoons NH_4^+ + OH^-\right)$ where $\left(NH_3\right)$ acts as a base and $\left(H2O\right)$ as an acid.

3. Lewis Concept
   i. Acids: Electron pair acceptors.
   ii. Bases: Electron pair donors.

Example: $\left(BF_3 + NH_3 \rightarrow F3BNH3\right)$, where $\left(BF_3\right)$ is a Lewis acid and $\left(NH_3\right)$ is a Lewis base (Shaik, 2007).

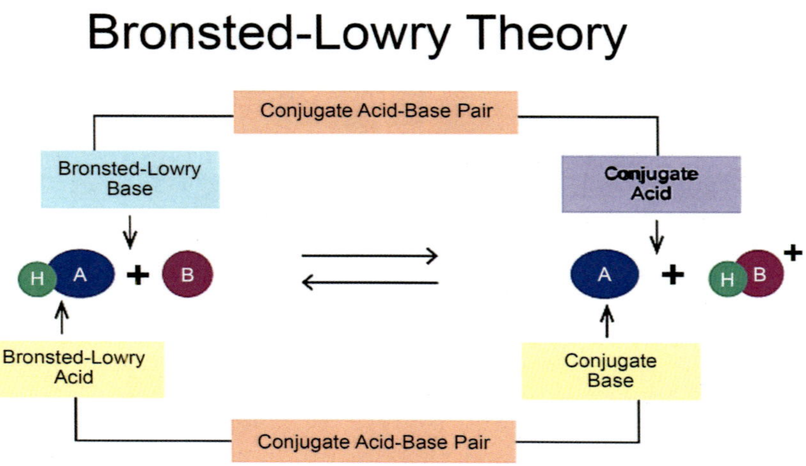

Figure 1.4. *Figure of Brønsted-Lowry concept.*

Source: Geeks for Geeks, Creative Commons License.

1.4.2. pH Scale and Its Measurement

The pH scale measures the acidity or basicity of a solution, defined as the negative logarithm of the hydrogen ion concentration:

$$pH = -\log\left[H^+\right]$$

a. Acidic solutions: $pH < 7$

b. Neutral solutions: $pH = 7$

c. Basic solutions: $pH > 7$

Measurement techniques include:

i. pH Meters: Provide precise readings by measuring the voltage difference between two electrodes.

ii. Indicator Papers: Change color based on the pH of the solution.

iii. Chemical Indicators: Substances like phenolphthalein and litmus, which show distinct color changes in acidic or basic environments (Radtke et al., 2021).

1.4.3. Acid-Base Titrations and Buffer Solutions

1. Acid-Base Titrations
   i. Purpose: To determine the concentration of an unknown acid or base.
   ii. Process: Involves adding a titrant of known concentration to the analyte until the equivalence point is reached, indicated by a color change or pH meter reading.

Example Calculation: If 25.0 mL of 0.1 M NaOH is needed to titrate 50.0 mL of HCl, the concentration of HCl can be found using (Karita and Kaneta, 2014):

$$M_1 \times V_1 = M_2 \times V_2$$

$$M_{HCl} = \frac{0.1 \times 25.0}{50.0} = 0.05 \text{ M}$$

2. Buffer Solutions
   i. Definition: Solutions that resist changes in pH upon the addition of small amounts of acids or bases.
   ii. Components: Consist of a weak acid and its conjugate base, or a weak base and its conjugate acid.
   iii. Henderson-Hasselbalch Equation: Used to calculate the pH of buffer solutions (Figure 1.5):

$$\text{pH} = \text{p}K_a + \log\left(\frac{[\text{A}^-]}{[\text{HA}]}\right)$$

where $(\text{p}K_a)$ is the acid dissociation constant, $([\text{A}^-])$ is the concentration of the conjugate base, and $([\text{HA}])$ is the concentration of the weak acid (Ni'mah et al., 2018).

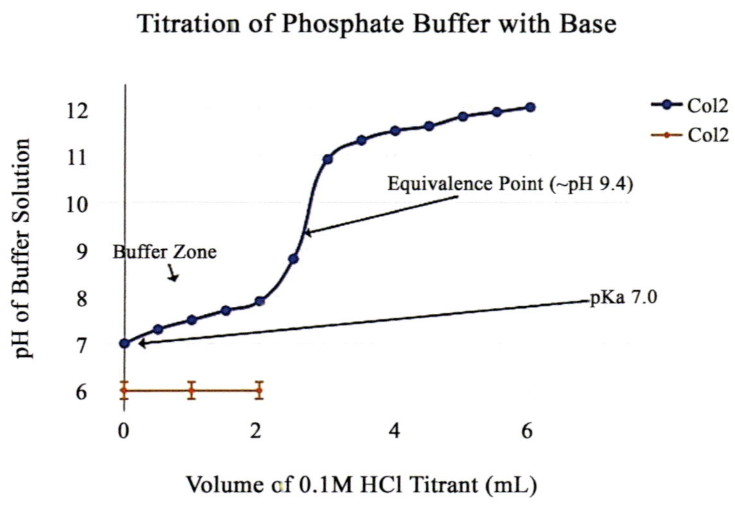

Figure 1.5. *Graphical representation buffer solution.*

Source: Lumen Learning, Creative Commons License.

1.4.4. Industrial Applications: Neutralization Reactions and pH Control

Acids, bases, and salts are fundamental to many industrial processes:

1. **Neutralization Reactions:** Principle: Involves the reaction of an acid and a base to form water and a salt.

Example: The reaction between hydrochloric acid and sodium hydroxide:

$$HCl + NaOH \rightarrow NaCl + H2O$$

Applications: Used in wastewater treatment to neutralize acidic or basic effluents, in the production of various chemicals, and in pharmaceuticals (Raczyńska et al., 2009).

2. pH Control

i. Importance: Maintaining the correct pH is crucial in processes like fermentation, electroplating, and chemical synthesis.

ii. Methods: Use of buffer systems, addition of acids or bases, and automated pH control systems to maintain optimal conditions for reactions and processes (Shi et al., 2009).

1.5. CHEMICAL REACTIONS AND STOICHIOMETRY

Chemical reactions are fundamental to both natural processes and engineered systems. This section covers the various types of chemical reactions, stoichiometry principles including limiting reactants and percent yield, and their applications in engineering such as combustion, corrosion, and electroplating (Chandrasegaran et al., 2009).

1.5.1. Types of Chemical Reactions

1. Synthesis Reactions
    i. Definition: Two or more reactants combine to form a single product.

    ii. General Form: $(A+B\rightarrow AB)$

Example: Formation of water from hydrogen and oxygen:

$$2H_2+O_2\rightarrow 2H2O$$

2. Decomposition Reactions
    i. Definition: A single compound breaks down into two or more simpler substances.

    ii. General Form: $(AB\rightarrow A+B)$

Example: Decomposition of calcium carbonate:

$$CaCO_3\rightarrow CaO+CO_2$$

3. Single Replacement Reactions
    i. Definition: An element replaces a similar element in a compound.

    ii. General Form: $(A-BC\rightarrow AC+B)$

Example: Reaction of zinc with hydrochloric acid:

$$Zn+2HCl\rightarrow ZnCl_2+H_2$$

4. Double Replacement Reactions

i. Definition: The ions of two compounds exchange places to form two new compounds.

ii. General Form: $\left(AB + CD \rightarrow AD + CB\right)$

Example: Reaction between sodium sulfate and barium chloride (Chen et al., 2013):

$$Na2SO_4 + BaCl_2 \rightarrow BaSO_4 + 2NaCl$$

1.5.2. Applications in Engineering

Applications in engineering span a diverse range of chemical processes, each pivotal in various industrial sectors. Combustion, for instance, epitomizes the rapid oxidation of substances like hydrocarbons such as methane, as expressed by the equation $\left(CH_4 + 2O_2 \rightarrow CO_2 + 2H2O\right)$. This process is indispensable in power generation within internal combustion engines and turbines, where understanding stoichiometry is essential for optimizing efficiency and controlling emissions.

Corrosion, on the other hand, entails the gradual degradation of metals due to chemical reactions, often involving oxidation. A prime example is the rusting of iron, represented by the equation $\left(4Fe + 3O_2 + 6H2O \rightarrow 4Fe(OH)_3\right)$. Engineers combat this phenomenon across infrastructure and machinery by employing protective coatings and inhibitors, underscoring the importance of understanding these chemical reactions in material durability and maintenance.

Lastly, electroplating exemplifies a method where electrical current is utilized to deposit a coherent metal coating onto an electrode surface. For instance, electroplating copper involves the reduction of $\left(Cu^{2+}\right)$ ions to metallic copper $\left(\left(Cu\right)\right)$, ensuring enhanced wear resistance, corrosion protection, and aesthetic appeal of manufactured metal components. These applications not only highlight the integration of chemical principles into engineering practices but also underscore their crucial role in enhancing functionality, longevity, and performance across industrial sectors.

Understanding the types of chemical reactions, stoichiometry, and their applications is essential for engineers to design efficient processes, optimize reactions, and develop innovative solutions in various fields (Liang et al., 2001).

1.6. ENERGY CHANGES IN CHEMICAL REACTIONS

Understanding the energy changes in chemical reactions is crucial for optimizing processes in engineering and industry. This section covers the nature of exothermic and endothermic reactions, the concepts of enthalpy, entropy, and Gibbs free energy, energy diagrams and activation energy, and the practical implications of these energy changes on energy efficiency in chemical processes.

1.6.1. Exothermic and Endothermic Reactions

Exothermic and endothermic reactions are fundamental concepts in chemistry, each playing distinct roles in various applications and industrial processes. Exothermic reactions, such as the combustion of methane represented by $\left(CH_4 + 2O_2 \rightarrow CO_2 + 2H2O + \text{energy}\right)$, release energy, typically in the form of heat, to the surroundings. This energy release is crucial in applications like power generation, where it fuels turbines and internal combustion engines, as well as in heating systems and industrial processes requiring heat as a byproduct.

In contrast, endothermic reactions absorb energy from the surroundings, exemplified by the decomposition of calcium carbonate: $\left(CaCO_3 + \text{energy} \rightarrow CaO + CO_2\right)$. These reactions typically require energy input to proceed, making them essential in applications such as refrigeration systems, where heat is absorbed from the surroundings to cool the environment. Additionally, endothermic reactions are utilized in thermal decomposition processes for materials and in chemical syntheses where controlled energy input is necessary to drive reactions towards desired products. Understanding the characteristics and applications of exothermic and endothermic reactions is pivotal in engineering and chemical industries for optimizing processes and achieving desired outcomes efficiently (Duff and Truhlar, 1975).

1.6.2. Enthalpy, Entropy, and Gibbs Free Energy

Enthalpy is a fundamental thermodynamic quantity that represents the total heat content of a system at constant pressure. It includes both internal energy and the energy required to make room for the system's volume under constant pressure. The change in enthalpy (Ä) during a chemical reaction is a measure of the heat absorbed or released by the system. For exothermic reactions, where heat is released to the surroundings, (ÄH) is negative because the enthalpy of the products (H_{products}) is lower than that of the reactants $(H_{\text{reactants}})$. Conversely, for endothermic reactions, where heat is absorbed from the surroundings, (ΔH) is positive.

Entropy is a measure of the randomness or disorder of a system. It quantifies the number of microstates available to a system at a specific energy level. The change in entropy (ΔS) during a reaction reflects the change in disorder. If the products of a reaction are more disordered than the reactants, (ΔS) is positive, indicating an increase in entropy. Conversely, if the products are more ordered, (ΔS) is negative. Entropy plays a crucial role in determining the spontaneity of a reaction: spontaneous processes tend to increase the overall entropy of the universe. Gibbs free energy is a thermodynamic potential that combines enthalpy and entropy to predict the spontaneity of a reaction under constant temperature and pressure conditions. The change in Gibbs free energy (ΔG) is given by the equation:

$$\Delta G = \Delta H - T\Delta S$$

where (T) is the temperature in Kelvin. A negative (ΔG) indicates that the reaction is spontaneous under the given conditions because the products have lower free energy than the reactants. Conversely, a positive (ΔG) indicates that the reaction is non-spontaneous without external intervention.

Understanding (ΔH),(ΔS), and (ΔG) is essential in chemical engineering for predicting reaction feasibility, optimizing processes, and designing efficient chemical systems (Alberty, 1969).

1.6.3. Energy Diagrams and Activation Energy

Energy diagrams provide a visual representation of the energy changes that occur during a chemical reaction. They typically include the energy levels of reactants, products, and intermediates, as well as the activation energy required for the reaction to proceed. In an exothermic reaction, where energy is released to the surroundings, the energy diagram slopes downward from the reactants to the products. This downward slope indicates that the products have lower energy than the reactants, with the difference represented as the heat released (ΔH). Conversely, for an endothermic reaction, which absorbs energy from the surroundings, the energy diagram slopes upward from the reactants to the products, illustrating that the products have higher energy than the reactants.

Activation energy is defined as the minimum amount of energy required for a chemical reaction to occur. It serves as a barrier that reactant molecules must overcome to form products. The rate of a reaction is greatly influenced by its activation energy; reactions with higher activation energies tend to proceed more slowly than those with lower activation energies. The relationship between activation energy, temperature, and reaction rate is quantitatively described by the Arrhenius equation:

$$k = Ae^{-\frac{E_a}{RT}}$$

where (k) is the rate constant, (A) is the pre-exponential factor, (R) is the gas constant, and (T) is the temperature in Kelvin. This equation highlights that as activation energy decreases, the rate constant (k) increases exponentially with temperature, thereby accelerating the reaction rate. Understanding activation energy is crucial in various fields of chemistry and engineering, as it influences the design and optimization of chemical processes, catalysis, and reaction mechanisms (Mao and Campbell, 2019).

1.7. CHEMICAL SAFETY AND HAZARD MANAGEMENT

Chemical safety and hazard management are paramount in engineering practices involving chemical substances. This section delves into the classification of chemical hazards based on flammability, reactivity, and toxicity, the importance of Safety Data Sheets (SDS) and labeling, guidelines for safe handling and storage of chemicals, and emergency procedures for spill management.

1.7.1. Chemical Hazard Classification

Chemical Hazard Classification involves categorizing substances based on their potential risks and characteristics, which are critical for ensuring safety in various industrial and environmental contexts.

1.7.1.1. Flammability

Flammability refers to a substance's propensity to ignite and burn when exposed to a flame or heat source. This characteristic is crucial in determining fire hazards and safety protocols in workplaces. Chemicals are classified into different categories based on their flammability, ranging from highly flammable substances like ethanol and gasoline to non-flammable ones such as water. Understanding a substance's flammability helps in implementing appropriate storage, handling, and firefighting measures to prevent fire accidents (Wang et al., 2018).

1.7.1.2. Reactivity

Reactivity describes how readily a substance undergoes chemical reactions, particularly those that may pose hazards such as explosions or releases of toxic gases. Chemicals are evaluated for their reactivity with water, air, and other substances to assess potential risks in handling and storage. Reactive substances include strong acids, alkali metals like sodium, and compounds containing peroxides. Proper identification and classification of reactive chemicals are essential for designing safe processes and preventing accidental releases or reactions that could endanger personnel and the environment.

1.7.1.3. Toxicity

Toxicity indicates a substance's ability to cause harm to living organisms upon exposure. Chemicals are categorized based on their acute toxicity (effects from short-term exposure) and chronic toxicity (effects from long-term exposure). Toxic substances encompass a wide range, from heavy metals such as lead and mercury to pesticides, industrial chemicals, and certain pharmaceuticals. Assessing toxicity levels helps in establishing safe exposure limits, implementing protective measures for workers, and minimizing environmental impacts through proper disposal and pollution control strategies (Khan et al., 2001).

1.7.2. Safety Data Sheets (SDS) and Labeling

Safety Data Sheets (SDS) are essential documents that provide comprehensive information about the properties, hazards, handling procedures, and emergency measures related to a specific chemical substance or mixture.

SDS typically include detailed sections on:

i. Chemical Composition: Lists ingredients and their concentrations in the product.
ii. Hazards Identification: Describes the potential health, physical, and environmental hazards associated with the chemical.
iii. Handling and Storage: Provides guidance on safe handling practices, storage conditions, and incompatibilities with other substances.
iv. Emergency Measures: Outlines procedures for responding to spills, fires, leaks, or exposure incidents (Eastlake et al., 2012).

Compliance with SDS is crucial for regulatory adherence and ensuring the safety of personnel in workplaces where chemicals are handled or stored. By providing accessible and up-to-date SDS for all chemicals used, organizations support informed decision-making, hazard communication, and emergency preparedness.

Chemical container labels play a pivotal role in communicating hazards and ensuring safe handling practices. Labels are mandated to include standardized information to effectively convey the risks associated with the chemical substance.

Key elements on chemical labels include:

i. Hazard Symbols: Pictograms that visually represent the nature of hazards, such as flammability, toxicity, or corrosivity.
ii. Signal Words: Indicate the severity of hazards, with "danger" used for more severe hazards and "warning" for less severe ones.
iii. Precautionary Statements: Provide guidance on safe handling practices, protective measures, and emergency procedures.

Labels follow international standards, such as the Globally Harmonized System of Classification and Labeling of Chemicals (GHS), which ensures uniformity and clarity in hazard communication globally. Standardized labeling helps workers quickly identify chemical hazards, understand associated risks, and implement appropriate safety measures to mitigate potential exposures and accidents (Pratt, 2002).

ACTIVITY 1.1.

Conduct a simple experiment to demonstrate the pH scale using household items like vinegar and baking soda. Measure and record pH values to observe acid-base reactions firsthand.

SUMMARY

- Chapter One of "Fundamentals of Engineering Chemistry" provides a comprehensive introduction to essential concepts in chemistry relevant to engineering disciplines. It begins by exploring emerging chemical technologies crucial for modern engineering applications. The chapter delves into fundamental principles such as SI units and conversions, significant figures in measurement precision, and the mole concept with its relation to Avogadro's number.
- Stoichiometry, including the balancing of chemical equations and calculations, is thoroughly covered, alongside discussions on states of matter and intermolecular forces. Solutions and their properties, including concentration units and colligative properties, are detailed to provide a solid foundation in chemical principles.
- Further topics include acids, bases, and salts, elucidating concepts from Arrhenius to Lewis theories, the pH scale, acid-base titrations, and their industrial applications. Types of chemical reactions and their relevance to engineering processes are also explored, along with the thermodynamic aspects of energy changes in chemical reactions, including exothermic and endothermic reactions, enthalpy, entropy, and Gibbs free energy.
- The chapter concludes with an overview of chemical safety and hazard management, discussing hazard classification, safety data sheets (SDS), and labeling practices essential for safe handling of chemicals in engineering environments.

REVIEW QUESTIONS

1. What are significant figures, and why are they important in chemical measurements?
2. Explain the concept of stoichiometry and its practical applications in chemical engineering.
3. Compare and contrast the Arrhenius, Brønsted-Lowry, and Lewis definitions of acids and bases.
4. How do colligative properties differ from other types of solution properties, and what are their engineering applications?

5. Describe the differences between exothermic and endothermic reactions, including their energy implications.
6. Why is safety data sheet (SDS) important in chemical hazard management? Provide examples of information typically found on an SDS.
7. How can engineers use the concepts of Gibbs free energy and activation energy to optimize chemical reactions?

MULTIPLE CHOICE QUESTIONS

1. Which concept in chemistry measures the amount of substance in a sample based on Avogadro's number?

a. Molarity
b. Molality
c. Mole
d. Mass

2. Colligative properties of solutions depend primarily on:

a. Temperature
b. Volume
c. Number of particles
d. Concentration

3. The pH scale measures:

a. Concentration of hydrogen ions
b. Concentration of hydroxide ions
c. Concentration of both hydrogen and hydroxide ions
d. Acidity only

4. Which type of chemical reaction releases heat to its surroundings?

a. Endothermic
b. Exothermic
c. Isobaric
d. Isometric

5. Safety Data Sheets (SDS) provide information about:

a. The chemical's ingredients and hazards
b. The chemical's cost and availability
c. The chemical's color and appearance
d. The chemical's shelf life and storage conditions

6. Enthalpy change in a chemical reaction is represented by the symbol:

a. ΔS
b. ΔH

c. ΔG

d. ΔE

7. Which type of energy diagram represents the energy change during a chemical reaction?

a. Potential energy diagram

b. Kinetic energy diagram

c. Activation energy diagram

d. Gibbs free energy diagram

Answers to Multiple Questions

1. (c) 2. (c) 3. (a) 4. (b) 5. (a)

6. (b) 7. (a)

REFERENCES

1. Alberty, R. A. (1969). Standard Gibbs free energy, enthalpy, and entropy changes as a function of pH and pMg for several reactions involving adenosine phosphates. *Journal of Biological Chemistry, 244*(12), 3290–3302.
2. Andersen, J. M., & Mack, J. (2017). Decouspling the Arrhenius equation via mechanochemistry. *Chemical Science, 8*(8), 5447–5453.
3. Bílek, M., Nodzyńska, M., Kopek-Putała, W., & Zimak-Piekarczyk, P. (2018). Balancing chemical equations using sandwich making computer simulation games as a supporting teaching method. *Problems of Education in the 21st Century, 76*(6), 779.
4. Bindel, T. H. (2021). Introducing high school students to the Avogadro number and the mole concept using discovery with calculations based on physical properties of elements, crystal structures, and 28Si spheres. *Journal of Chemical Education, 98*(3), 790–795.
5. Bularzik, J. (2007). The penny experiment revisited: An illustration of significant figures, accuracy, precision, and data analysis. *Journal of Chemical Education, 84*(9), 1456.
6. Carpenter, J., Badve, M., Rajoriya, S., George, S., Saharan, V. K., & Pandit, A. B. (2017). Hydrodynamic cavitation: An emerging technology for the intensification of various chemical and physical processes in a chemical process industry. *Reviews in Chemical Engineering, 33*(5), 433–468.
7. Chandrasegaran, A. L., Treagust, D. F., Waldrip, B. G., & Chandrasegaran, A. (2009). Students' dilemmas in reaction stoichiometry problem solving: Deducing the limiting reagent in chemical reactions. *Chemistry Education Research and Practice, 10*(1), 14–23.
8. Chen, L., Kuai, L., Yu, X., Li, W., & Geng, B. (2013). Advanced catalytic performance of Au–Pt double-walled nanotubes and their fabrication through galvanic replacement reaction. *Chemistry–A European Journal, 19*(35), 11753–11758.

9. Dow, W. G. (1952). *Fundamentals of Engineering Electronics* (Vol. 2, pp. 186). Wiley.

10. Duff, J. W., & Truhlar, D. G. (1975). Effect of curvature of the reaction path on dynamic effects in endothermic chemical reactions and product energies in exothermic reactions. *The Journal of Chemical Physics, 62*(6), 2477–2491.

11. Eastlake, A., Hodson, L., Geraci, C., & Crawford, C. (2012). A critical evaluation of material safety data sheets (MSDSs) for engineered nanomaterials. *Journal of Chemical Health & Safety, 19*(5), 1–8.

12. Fahlman, B. D. (2023). Solid-state chemistry. In *Materials Chemistry* (Vol. 2, pp. 31–189). Springer International Publishing.

13. Flanagan, R. J. (1995). SI units—common sense not dogma is needed. *British Journal of Clinical Pharmacology, 39*(6), 589.

14. Ford, E. N., & Gilbert, Y. V. (2013). Displacement between orders of magnitude method for SI unit conversion. *Journal of Chemical Education, 90*(1), 134–136.

15. Gao, R., & Lloyd, J. (2020). Precision and accuracy: Knowledge transformation through conceptual learning and inquiry-based practices in introductory and advanced chemistry laboratories. *Journal of Chemical Education, 97*(2), 368–373.

16. Gillespie, L. J. (1936). Methods for the thermodynamic correlation of high pressure gas equilibria with the properties of pure gases. *Chemical Reviews, 18*(2), 359–372.

17. Haller, P. D., Bradley, L. C., & Gupta, M. (2013). Effect of surface tension, viscosity, and process conditions on polymer morphology deposited at the liquid–vapor interface. *Langmuir, 29*(37), 11640–11645.

18. Karita, S., & Kaneta, T. (2014). Acid–base titrations using microfluidic paper-based analytical devices. *Analytical Chemistry, 86*(24), 12108–12114.

19. Khan, F. I., Husain, T., & Abbasi, S. A. (2001). Safety weighted hazard index (SWeHI): A new, user-friendly tool for swift yet comprehensive hazard identification and safety evaluation in chemical process industries. *Process Safety and Environmental Protection, 79*(2), 65–80.

20. Korbag, I., & Saleh, S. M. (2017). Designing simple equations to fully understand the conversions among common concentration units: An overview. *Libyan Journal of Science & Technology, 6*(1), 654–660.

21. Larson, D. C., Weinberger, C. B., Lawley, A., Thomas, D. H., & Moore, T. W. (1994). Fundamentals of engineering energy. *Journal of Engineering Education, 83*(4), 325–330.

22. Li, X., & Wang, C. H. (2020). 2017 PV Danckwerts Memorial Lecture special issue editorial: Advances in emerging technologies of chemical engineering towards sustainable energy and environment: Solar and biomass. *Chemical Engineering Science, 215*, 115384.

23. Liang, Y. Z., Fang, K. T., & Xu, Q. S. (2001). Uniform design and its applications in chemistry and chemical engineering. *Chemometrics and Intelligent Laboratory Systems, 58*(1), 43–57.

24. Manning, G. S. (1969). Limiting laws and counterion condensation in polyelectrolyte solutions I. Colligative properties. *The Journal of Chemical Physics, 51*(3), 924–933.

25. Mao, Z., & Campbell, C. T. (2019). Apparent activation energies in complex reaction mechanisms: A simple relationship via degrees of rate control. *ACS Catalysis, 9*(10), 9465–9473.

26. Ni'mah, F., Ibnu, S., & Rahayu, S. (2018). How guided inquiry and coupled inquiry influence students' attitude toward chemistry in buffer solution and solubility topics. In *AIP Conference Proceedings* (Vol. 2049, pp. 54–58). AIP Publishing.

27. Niaz, M., & Montes, L. A. (2012). Understanding stoichiometry: Towards a history and philosophy of chemistry. *Educación Química, 23*, 290–297.

28. Powsner, E. R. (1984). SI quantities and units for American medicine. *JAMA, 252*(13), 1737–1741.

29. Pratt, I. S. (2002). Global harmonization of classification and labelling of hazardous chemicals. *Toxicology Letters, 128*(1–3), 5–15.

30. Raczyńska, E. D., Gal, J. F., Maria, P. C., & Szel⊠g, M. (2009). Proton transfer chemistry in the gas phase: Is a spontaneous 'neutralization' reaction a myth or a reality? *Croatica Chemica Acta, 82*(1), 87–103.

31. Radtke, V., Stoica, D., Leito, I., Camões, F., Krossing, I., Anes, B., & Lawrence, N. (2021). A unified pH scale for all solvents: Part I—intention and reasoning (IUPAC Technical Report). *Pure and Applied Chemistry, 93*(9), 1049–1060.

32. Romine, W. L., Todd, A. N., & Clark, T. B. (2016). How do undergraduate students conceptualize acid–base chemistry? Measurement of a concept progression. *Science Education, 100*(6), 1150–1183.

33. Rubin, A. E., Tummala, S., Both, D. A., Wang, C., & Delaney, E. J. (2006). Emerging technologies supporting chemical process R&D and their increasing impact on productivity in the pharmaceutical industry. *Chemical Reviews, 106*(7), 2794–2810.

34. Scatchard, G., Coleman, J. S., & Shen, A. L. (1957). Physical chemistry of protein solutions. VII. The binding of some small anions to serum albumin. *Journal of the American Chemical Society, 79*(1), 12–20.

35. Shaik, S. (2007). The Lewis legacy: The chemical bond—A territory and heartland of chemistry. *Journal of Computational Chemistry, 28*(1), 51–61.

36. Shi, D., Xu, Y., & Morel, F. M. M. (2009). Effects of the pH/pCO_2 control method on medium chemistry and phytoplankton growth. *Biogeosciences, 6*(7), 1199–1207.

37. Tarhan, L., Ayar-Kayali, H., Urek, R. O., & Acar, B. (2008). Problem-based learning in 9th grade chemistry class: 'Intermolecular forces.' *Research in Science Education, 38*, 285–300.

38. Tóth, Z., & Sebestyén, A. (2009). Relationship between students' knowledge structure and problem-solving strategy in stoichiometric problems based on the chemical equation. *International Journal of Physics and Chemistry Education, 1*(1), 8–20.

39. Wang, B., Wu, C., Reniers, G., Huang, L., Kang, L., & Zhang, L. (2018). The future of hazardous chemical safety in China: Opportunities, problems, challenges and tasks. *Science of the Total Environment, 643*, 1–11.

40. Widom, B. (1967). Intermolecular forces and the nature of the liquid state: Liquids reflect in their bulk properties the attractions and repulsions of their constituent molecules. *Science, 157*(3787), 375–382.

41. Woods, D. R., & Sawchuk, R. J. (1993). Fundamentals of chemical engineering. *Chemical Engineering Education, 27*(2), 80–85.

42. Yang, Y., Luo, M., Zhang, W., Sun, Y., Chen, X., & Guo, S. (2018). Metal surface and interface energy electrocatalysis: Fundamentals, performance engineering, and opportunities. *Chem, 4*(9), 2054–2083.

43. Zhang, S., Shao, T., Bekaroglu, S. S. K., & Karanfil, T. (2010). Adsorption of synthetic organic chemicals by carbon nanotubes: Effects of background solution chemistry. *Water Research, 44*(6), 2067–2074.

CHAPTER 2

Atomic Structure and Bonding

LEARNING OBJECTIVES

After studying this chapter, you will be able to:

1. Understand the basics of atomic structure including the roles of protons, neutrons, and electrons.
2. Explain Rutherford's Gold-foil experiment and its significance in the development of atomic theory.
3. Describe Bohr's atomic model and its postulates.
4. Differentiate between atomic number (Z) and mass number (A) and their significance in identifying elements.
5. Define isotopes, isobars, and isotones, and their properties.
6. Discuss the Aufbau principle and its application in determining electron configurations.
7. Explain different types of chemical bonding including ionic, covalent, and coordinate bonds.

KEY TERMS FROM THIS CHAPTER

Atomic number (Z)
Covalent bonding
Isotones
Mass number (A)
Aufbau principle
Isobars
Isotopes

INTRODUCTION

Understanding atomic structure and bonding in detail is necessary to comprehend how matter behaves at the tiny level. The building blocks of all matter, atoms are made up of a core nucleus that is made up of neutral neutrons and positively charged protons (Xu et al., 2009). These are covered by negatively charged electrons that occupy regions known as orbitals; these orbitals are critical in defining an atom's chemical properties (Siegel et al., 2002).

The positioning of these electrons, regulated by factors such as the Aufbau Principle and electronic configuration, determines the manner in which atoms engage in chemical bonding to create compounds. Ionic bonding is the process of moving electrons from one atom to another, creating ions that are drawn to one another by electrostatic forces (Wang et al., 1995).

Contrasting with other types of bonding, covalent bonding occurs when atoms share electron pairs, resulting in the formation of stable molecules. The understanding of bonding mechanisms is crucial for explaining a wide range of chemical phenomena, spanning from the creation of basic diatomic molecules to detailed macromolecular structures (Abdoolatiff and Narod, 2009).

The chapter also covers Lewis symbols, the octet rule, and exceptions that offer valuable insights into the stability and reactivity of various chemical species. Understanding these principles is essential for materials science and biochemistry because they explain the underlying physical and chemical aspects of our universe (Ren and Meng, 2006).

2.1. BASICS OF ATOMIC STRUCTURE

The fundamental principle of Dalton's atomic theory is that matter is made up of indivisible particles called "atoms" that cannot be further divided. According to recent studies, atoms are truly divisible and have a complicated structure made up of many different subatomic particles, including quarks, electrons, neutrons, protons, antiprotons, mesons, leptons, positrons, neutrinos, and antineutrinos (Grant and Quiney, 1988).

The fundamental particles of all matter are the subatomic particles neutrons, electrons, and protons.

i. **Electron:** The electron is a fundamental subatomic particle having a charge of $-1.602x10^{-19}$ Coulomb and minuscule mass of $9.11x1^{-31}$ kg.

ii. **Proton:** A proton is a basic subatomic particle a charge of $+1.602X10^{-19}$ Coulomb and with a mass of $1.672x10^{-27}$ kg.

iii. **Neutron:** A neutron is a basic subatomic particle with no electric charge and a mass of $1.672x10^{-27}$ kg (Jeffries et al., 2020).

Fundamental Particle	**Charge**	**Mass**	**Relative Charge**
Electron	$-1.602x10^{-19}$	$9.11x1^{-31}$	−1
Proton	$+1.602X10^{-19}$	$1.672x10^{-27}$	+1
Neutron	No Charge	$1.672x10^{-27}$	0

Remember

Remember the basics of atomic structure are foundational to understanding chemical bonding and the behavior of matter at the atomic level.

2.2. RUTHERFORD'S GOLD-FOIL EXPERIMENT

In 1911, E. Rutherford first offered important insights into the highly precise representation of an atom. During the experiment, a stream of α-particles (He^{+2} ions) was directed towards a thin sheet of gold foil. This was achieved by using a radioactive substance like Uranium. A circular display made of zinc sulfide (ZnS) was placed behind the gold foil to record the marks left by the α-particles (Leone et al., 2018).

2.2.1. Observations and Conclusions

Rutherford made a key observation through the α-scattering experiment:

1. Most of the α-particles did not change direction at all and passed straight through the gold foil. It is clear that a substantial part of an atom is made up of empty space.
2. Some α-particles experienced significant deflection from their normal pathways. There is evidence of a substantial positively charged body residing within the atom. The nucleus is a heavy, positively charged body.
3. Only a small fraction (0.01%) of α-particles exhibited an extraordinary deviation from their initial paths, being deflected by nearly 1800°. One can observe that the nucleus is incredibly tiny in size. Atomic nuclei are generally quite small, with a size of approximately 10^{-13} cm (Buist et al., 2004).

2.2.2. Rutherford's Atomic Model

The atomic model developed by Rutherford was a result of the observations made during the α-scattering experiment:

1. An atom is composed of two components: the nucleus and the extranuclear region.
2. Each atom is composed of an extremely small yet massive positively charged core, known as the nucleus.
3. An atom's nucleus contains the majority of its mass.
4. The speed at which electrons revolve around the nucleus is comparable to that of planets orbiting the sun. As a result, planetary electrons are a common description for electrons.

5. Atomic structure and bonding require a fine balance between electrostatic and centrifugal forces. This effect is caused by a repulsive electrostatic force between negatively charged electrons and positively charged nuclei. This force functions as a barrier, keeping electrons from collapsing into the nucleus (Yock, 2021).

Did you know?

Did you know that Rutherford's gold-foil experiment demonstrated that most of the atom's mass is concentrated in a tiny, positively charged nucleus? This pivotal experiment provided evidence for the nuclear model of the atom.

2.2.3. Drawbacks of Rutherford's Atomic Model

The Rutherford atomic model (Figure 2.1) has several drawbacks:

1. Stability of Atom: The hypothesis does not provide an explanation for the atoms stability. According to Clark Maxwell's law of electrodynamics, energy is always released by a charged particle orbiting another charged particle. The spinning electron is attracted toward the nucleus as its energy drops, gradually following a spiral path until it gets there. Sadly, this never happens.
2. No information about the precise energy and velocity of the rotating electrons is provided by the model.
3. The theory does not provide a satisfactory explanation for atomic spectra.
4. The explanation for chemical combinations is not provided (Niaz, 1998).

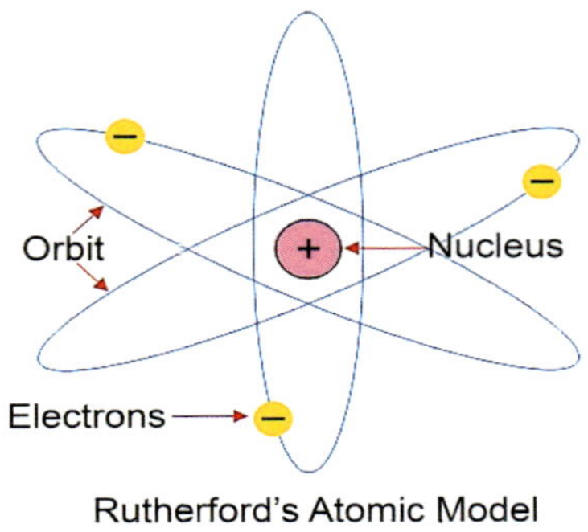

Figure 2.1. Figure of Rutherford atomic model.

Source: Socratic, Creative Common License.

2.3. BOHR'S ATOMIC MODEL

Niels Bohr, a Dutch physicist, presented an almost accurate depiction of the atomic model in 1913. Planck's Quantum Theory and the idea of energy quantization serve as the foundation for the Bohr Atomic Model (McKagan et al., 2008).

2.3.1. Postulates of Bohr's Atomic Model

Following are the fundamental principles of Bohr's Model:

1. Each atom is composed of a centrally located, heavily positively charged nucleus, around which electrons circle in specific, authorized circular routes referred to as "shells," "orbits," or "stationary states."
2. The first, second, third, and fourth shells are represented by the shells or stationary states, which are designated as K, L, M, N, O, and so forth
3. Each shell possesses a certain and fixed amount of energy. Therefore, the shells are sometimes referred to as energy levels (Bohr, 1915).
4. The energy levels of different shells are compared using the formula:

$$E_n = -\frac{2\pi^2 me^4 z^2}{n^2 h^2}$$

where,

m = The charge of an electron equals its mass, e.

z = Atomic number.

h = Planck's constant = 6.625 x 10^{-27} erg sec.

n = Shell number (1, 2, 3) or Principal quantum number.

$$E_n = -\frac{1312}{n^2} KJ / Mol,$$ for Hydrogen atom

1. The energy level rises as one move from lower to higher electron shells, reaching 0 for the shell located at an infinite distance from the nucleus.

2. The energy levels have non-uniform spacing (Darrigol, 2013).
3. Despite the availability of several circular pathways, electrons are only permitted to orbit around the nucleus by circular pathways with an angular momentum value (mvr) that is a whole integer multiple of h/2π.

$$mvr = n\frac{h/}{2\pi}, \text{ where } n = 1, 2, 3, ...$$

Here,

where m is the electron's mass,

v is its tangential velocity while it is rotating, and

r is the orbit's radius.

h is the Planck constant.

In the ground state of an atom, electrons orbit continuously in their respective paths without experiencing any energy loss.

1. An atom's electrons absorb energy from an external energy source to move from lower energy levels to higher energy levels.
2. Electrons in excited states display instability and undergo transitions to lower energy ground states, resulting in the emission of energy. The energy that is emitted appears as light, which is the origin of spectral lines (Hughes, 1990).

2.3.2. Failures of Bohr's Atomic Model

The failures of Bohr's atomic model are as follows:

1. In Bohr's atomic model (Figure 2.2), electrons move along a two-dimensional circular trajectory. Recent scientific discoveries, including Heisenberg's Uncertainty Principle, have shown that electrons follow 3-dimensional paths called orbitals.
2. It does not provide an explanation for the spectra observed in organisms with multiple electrons.
3. It does not provide an explanation for the comparative strengths of spectral lines.
4. There is no adequate explanation for how a magnetic field (Zeeman Effect) and electric field (Stark Effect) split spectral lines.
5. It does not provide an explanation for the cause of chemical combinations (Schwarz, 2013).

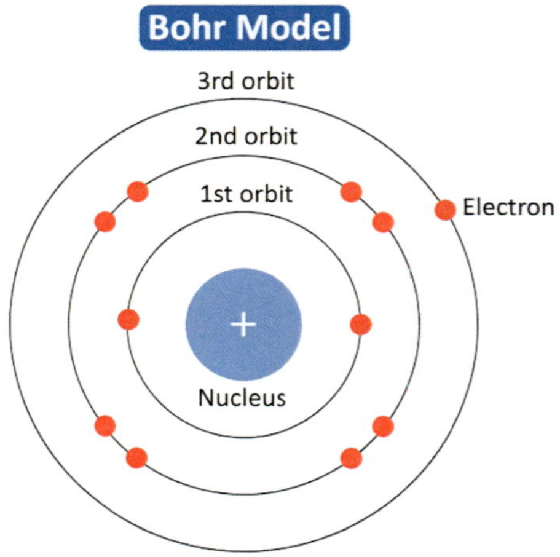

Figure 2.2. *Schematic of Bohr atomic model.*

Source: Chemistry learner, Creative Common License.

2.3.3. Bohr-Bury Scheme

The Bohr-Burry scheme explores an organization of electrons within different shells. There are several postulates of the scheme:

1. In atomic structure, a shell has the capacity to hold a maximum of $2n^2$ electrons. When considering the shell, it is important to take into account the value of n.

Shell	N	Maximum No. of Electrons = 2n2
K	1	2 X 12 = 2
L	2	2 x 22 = 8
M	3	2 x 33 = 18
N	4	2 X 44 = 32, and so on.

2. The valence shell, which is the outermost shell, may hold a maximum of eight electrons.
3. The second-to-last shell, also known as the penultimate shell, has a maximum capacity of 18 electrons.
4. In certain cases, a higher orbit may begin to fill before the lower orbit has been fully filled (Kragh, 1979).

Here are the electron arrangements in various orbits of certain elements, as per the Bohr-Bury scheme (Figure 2.3).

Element		Electrons in		
	L-Shell	K-Shell	M-Shell	N-Shell
$_{1}H$		1		
$_{2}He$		2		
$_{3}Li$	1	2		
$_{4}Be$	2	2		
$_{5}B$	3	2		
$_{10}Ne$	8	2		
$_{11}Na$	8	2	1	
$_{12}Mg$	8	2	2	
$_{13}Al$	8	2	3	
$_{18}Ar$	8	2	8	
19K	8	2	8	1
$_{20}Ca$	8	2	8	2
$_{21}Sc$	8	2	9	2

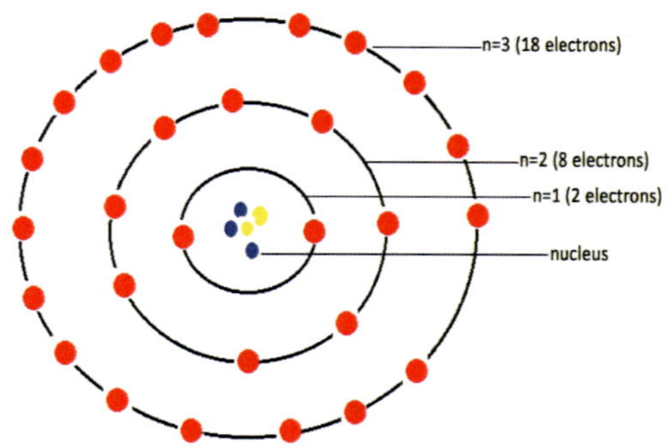

Figure 2.3. *Diagram of Bohr Bury scheme.*

Source: Vedantu, Creative Common License.

2.4. MASS NUMBER (A) AND ATOMIC NUMBER(Z)

Following section explains Mass number and atomic number.

2.4.1. Atomic Number(Z)

The total number of protons found in a single atom is known as the atomic number.

Atomic Number = No. of Protons (p)

For examples (Karol et al., 2016):

Element	No of Protons (p)	Atomic No. (Z)
Nitrogen (N)	7	7
Carbon (C)	6	6
Oxygen (O)	8	8

2.4.2. Mass Number (A)

The total number of neutrons and protons in an atom is known as the element's mass number (Xu et al., 2013).

Tip

The atomic number (Z) represents the number of protons in an atom's nucleus, determining its chemical identity and place in the periodic table. It is crucial for understanding atomic structure and chemical properties.

Mass Number (A) = No. of Protons (p) + No. of Neutrons (n)

Element	No of Neutrons (p)	No. of Protons (n)	Mass No. (A)
Nitrogen (N)	7	7	14
Carbon (C)	6	6	12
Sodium (Na)	12	11	23

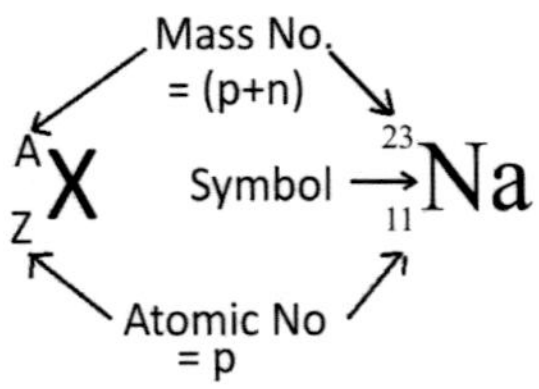

Note: $A - ne^{-} \rightarrow A^{n+}$ Example: $Mg - 2e^{-} \rightarrow Mg^{2+}$

$A + ne^{-} \rightarrow A^{n-}$ Example: $N + 3e^{-} \rightarrow N^{3-}$

2.5. ISOTOPES

Isotopes are atoms of the same element that possess different mass numbers while maintaining the same atomic number (Dawson et al., 2002).

Examples: Hydrogen has 3 isotopes: Protium Here is the chemical symbol for Carbon-12 written in Unicode: (^1H), Deuterium (^2H) and Tritium (^3H) Similarly, $^{35}Cl_{17}$ and $^{37}Cl_{17}$.

2.5.1. Isotopes Properties

1. These are identical atoms of the same chemical element.
2. These elements possess identical atomic numbers but exhibit distinct mass numbers.
3. They have same number electrons and protons.
4. They vary in the number of neutrons they contain.
5. They exhibit analogous chemical characteristics.
6. These substances exhibit distinct physical characteristics, including melting point, boiling point, density, viscosity, and so on (Goncharuk et al., 2011).

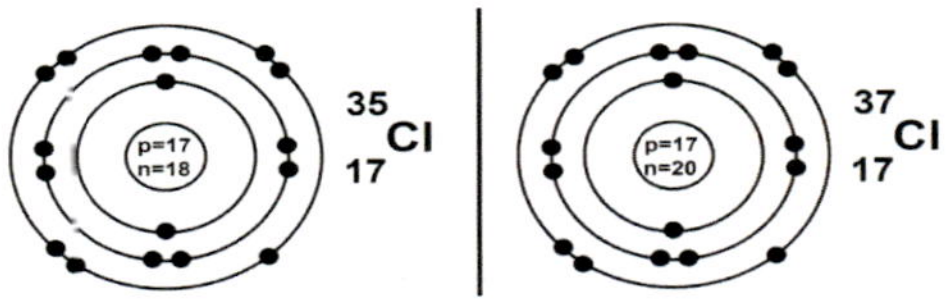

Figure 2.4. *Illustration of isotopes of chlorine.*

Source: Knowledge Boat, Creative Common License.

2.5.2. Isobars

Atoms of various elements with different atomic numbers but the same mass number are called isobars.

Examples: $^{40}Cl_{18}$ and $^{18}Cl_{20}$.

2.5.2.1. Properties of Isobars

1 These are atoms belonging to distinct chemical elements.
2 These elements have distinct atomic numbers but same mass numbers.
3 These possess varying numbers of protons and electrons.
4 These possess distinct chemical characteristics.
5 These substances exhibit distinct physical characteristics, including melting point, boiling point, density, viscosity, and so on (Chiang et al., 2002).

2.5.3. Isotones

Atoms of elements with an identical number of neutrons are called isotones.

Examples:

$^{23}_{11}Na$ and $^{24}_{12}Mg$ $^{27}_{13}Al$ and $^{28}_{14}Si$ $^{31}_{15}P$ and $^{32}_{16}S$

Note: The particles found within the nucleus, namely protons and neutrons, are collectively referred to as nucleons (Kleitman and Markowsky, 1975).

2.6. AUFBAU PRINCIPLE

The term "Aufbau" denotes the process of constructing or assembling something. This principle states the process by which electrons occupy subshells. The Aufbau principle can be defined as the arrangement of electrons in sub-shells according to their increasing energy levels. Electrons are initially filled in the sub-shell with the lowest energy, followed by the filling of sub-shells with increasing energies. The energy levels of different sub-shells can be compared using the (n+ l) rule (Guo et al., 2019).

2.6.1. The (n + l) Rule

1. The sub-shell that fills first and has a lower (n+*l*) value has less energy.
2. The sub-shell with the lower value of 'n' has less energy and is filled before the other sub-shell if the sum of the (n+*l*) for the two sub-shells is equal.

Applying the (n+*l*) rule, we will now compare the energy levels of different sub-shells.

Sub-shell→	**1s**	**2s**	**2p**	**3s**	**3p**	**3d**	**4s**	**4p**	**4d**	**4f**	**5s**	**5p**
(n+*l*)	1+0 = 1	2+0 = 2	2+1 = 3	3+0 = 3	3+1 = 4	3+2 = 5	4+0 = 4	4+1 = 5	4+2 = 6	4+3 = 7	5+0 = 5	5+1 = 6

Therefore, the sub-shells are arranged in increasing order of energy content (Metz et al., 2021).

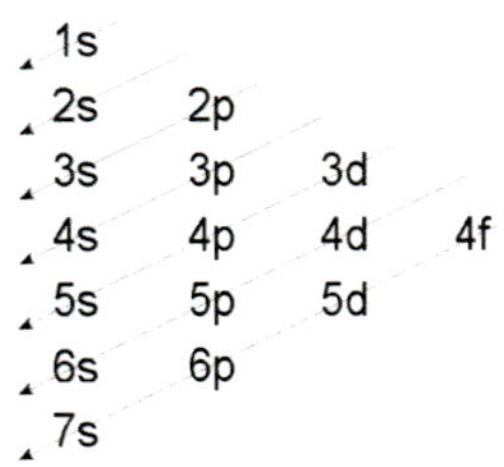

1s < 2s < 2p < 3s < 3p <4s <3d < 4p <..

Figure 2.5. *Illustration of Aufbau principle.*

Source: 88Guru, Creative Common License.

2.7. ELECTRONIC CONFIGURATION

The atom's electrons arrangements in various subshells or orbitals in ascending order of energy content is known as electronic configuration. The following list includes the electrical setups of several elements (Brik and Srivastava, 2017).

Elements	Electronic Configurations	Elements	Electronic Configurations
$_1$H	$1s^1$	16S	$1s^2\ 2s^2 2p^6\ 3s^2 3p^4$
$_2$He	$1s^2$	$_{17}$Cl	$1s^2\ 2s^2 2p^6\ 3s^2 3p^5$
$_3$Li	$1s^2\ 2s^1$	$_{18}$Ar	$1s^2\ 2s^2 2p^6\ 3s^2 3p^6$
$_4$Be	$1s^2\ 2s^2$	19K	$1s^2\ 2s^2 2p^6\ 3s^2 3p^6 4s^1$
$_5$B	$1s^2\ 2s^2 2p^1$	$_{20}$Ca	$1s^2\ 2s^2 2p^6\ 3s^2 3p^6 4s^2$
$_6$C	$1s^2\ 2s^2 2p^2$	$_{21}$Sc	$1s^2\ 2s^2 2p^6\ 3s^2 3p^6 4s^2 3d^{1}$
$_7$N	$1s^2\ 2s^2 2p^3$	$_{22}$Ti	$1s^2\ 2s^2 2p^6\ 3s^2 3p^6 4s^2 3d^{2}$
$_8$O	$1s^2\ 2s^2 2p^4$	23V	$1s^2\ 2s^2 2p^6\ 3s^2 3p^6 4s^2 3d^{3}$
$_9$F	$1s^2\ 2s^2 2p^5$	$_{24}$Cr	$1s^2\ 2s^2 2p^6\ 3s^2 3p^6 4s^1 3d^{5}$
$_{10}$Ne	$1s^2\ 2s^2 2p^6$	$_{25}$Mn	$1s^2\ 2s^2 2p^6\ 3s^2 3p^6 4s^2 3d^{5}$
$_{11}$Na	$1s^2\ 2s^2 2p^6\ 3s^1$	$_{26}$Fe	$1s^2\ 2s^2 2p^6\ 3s^2 3p^6 4s^2 3d^{6}$
$_{12}$Mg	$1s^2\ 2s^2 2p^6\ 3s^2$	$_{27}$Co	$1s^2\ 2s^2 2p^6\ 3s^2 3p^6 4s^2 3d^{7}$
$_{13}$Al	$1s^2\ 2s^2 2p^6\ 3s^2 3p^1$	$_{28}$Ni	$1s^2\ 2s^2 2p^6\ 3s^2 3p^6 4s^2 3d^{8}$
$_{14}$Si	$1s^2\ 2s^2 2p^6\ 3s^2 3p^2$	$_{29}$Cu	$1s^2\ 2s^2 2p^6\ 3s^2 3p^6 4s^1 3d^{10}$
15P	$1s^2\ 2s^2 2p^6\ 3s^2 3p^3$	$_{30}$Zn	$1s^2\ 2s^2 2p^6\ 3s^2 3p^6 4s^2 3d^{10}$

2.7.1. Exceptional Electronic Configuration

Certain elements exhibit remarkable electrical configurations, such as ⊠Cr and ⊠Cu

The ideal electrical arrangements for ⊠Cr and ⊠Cu are as follows:

$_{24}$Cr = [Ar]$4s^2 3d^{4}$ & $_{29}$Cu = [Ar]$4s^2 3d^{9}$, respectively. However, the actual electronic configurations are

$_{24}$Cr = [Ar]$4s^1 3d^{5}$ & $_{29}$Cu = [Ar]$4s^1 3d^{10}$

Because of orbital symmetry and exchange energy, full-filled, and half-filled orbitals are more stable, which accounts for a unique electronic configuration (Choi et al., 2018).

2.8. CHEMICAL BONDING

Chemical bonds are the powerful forces that keep the atoms of a molecule or an ion connected to each other.

Chemical bonding (Figure 2.6) can be divided into the following categories based on the way the bonds are formed (transfer or sharing of electrons):

1. Electrovalent bonding or Ionic Bonding.
2. Covalent bonding.
3. Dative Bonding or Co-ordinate bonding.
4. Hydrogen bonding.
5. Metallic bonding (Goddard and Harding, 1978).

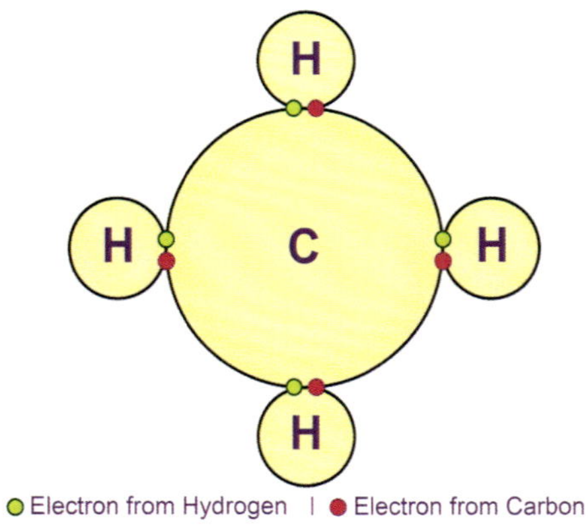

Figure 2.6. *Example of chemical bonding.*

Source: Byjus, Creative Commons License.

2.8.1. Ionic or Electrovalent Bonding

Ionic compounds, often known as salts, are compounds made up of ions that are bound together by ionic bonds. Anions and cations with opposing charges are drawn to each other by electrostatic forces. The characteristics of ionic compounds provide insight into the fundamental nature of ionic bonding. Ionic solids possess a crystalline arrangement and display rigidity and brittleness. Additionally, they have high melting and boiling temperatures, indicating the presence of exceptionally strong ionic connections (Verma, 2015).

Ionic solids have low electrical conductivity due to the strong ionic connections that hinder the movement of ions within the solid state. Nevertheless, the majority of ionic materials quickly dissolve in water. After undergoing dissolution or melting, ionic compounds exhibit exceptional conductivity for both electricity and heat due to the unrestricted movement of ions.

Neutral atoms and the ions they are associated with have entirely distinct atomic and chemical characteristics. Sodium metal is a flexible and lustrous metal composed of sodium atoms. When it comes into touch with water, it reacts explosively and burns vigorously in the air (Gordy,1946).

Chlorine gas, also referred to as Cl_2, is a yellow-green gas consisting of chlorine atoms. It exhibits a high level of corrosiveness towards the majority of metals and poses significant harm to both plants and animals.

Chlorine and sodium react violently to form sodium chloride, also known as common table salt. Sodium chloride is a white, crystalline material. It is made up of chloride anions and sodium cations. The characteristics of the molecule made up of these ions are completely different from those of the elements chlorine and sodium. Chlorine is harmful when sodium atoms react forcefully with water, yet sodium chloride just dissolves in it and is essential for life (Aman, 2020).

Due to its flexibility, sodium must be kept in mineral oil to avoid exposure to moisture or air. (b) A gas with a little yellow-green hue is chlorine. (c) These ingredients combine chemically to form sodium chloride, also referred to as table salt and appearing as white crystals.

2.8.1.1. The Ionic Compounds Formation

Binary ionic compounds consist of two components: a nonmetal that generates the anions and a metal that generates the cations. One example of a binary ionic compound is sodium chloride (NaCl). Considering the periodic properties of the elements, the production of such compounds becomes evident. Several metallic elements have high electron emission rates and possess comparatively low ionization potentials.

On the periodic table, these elements are located to the left of a period or in the lower half of a group. Because of their strong electron affinities, nonmetal atoms can easily receive electrons that are emitted by metal atoms. This technique enables nonmetal atoms to achieve a full complement of electrons in their outermost energy levels. The periodic table's upper right section contains the nonmetallic elements (Hautier et al., 2011).

Ionic compounds must have equal positive and negative charges on their cations and anions to be electrically neutral. An ionic compound's formula gives the most basic proportion of ions needed to generate equal positive and negative charges. Aluminum oxide (Al_2O_3) consists of two aluminum cations (Al^{3+}) and three oxide anions (O^{2-}), resulting in a charge balance of $(2 \times +3) + (3 \times -2) = 0$.

It is crucial to remember that an ionic molecule's formula does not represent its ions' physical configuration. Because sodium and chloride ions do not form a single ionic connection, sodium chloride (NaCl) is not a "molecule." Due to the isotropic nature of attractive interactions between ions, nearby ions with opposite charges will be equally attracted to each other (Bidlingmeyer, 1980).

As a result, the ions organize themselves into a solid lattice structure that forms a strong bond. For example, sodium chloride consists of a balanced combination of Cl– anions and Na+ cations arranged in a regular pattern.

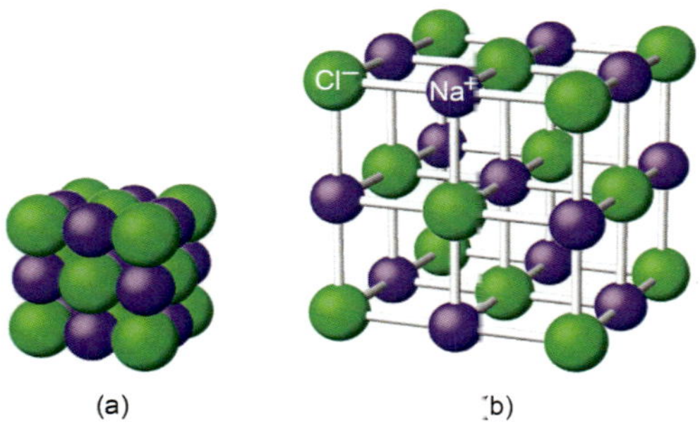

Figure 2.7. *Figure of sodium chloride, or regular table salt, has its atoms organized to (a) maximize the interaction of opposing charges. Ions of sodium and chloride are represented by the smaller and larger spheres, respectively. In the expanded view (b), It is easier to see the geometry. Every ion is "bonded" to every other ion in its immediate vicinity—six.*

Strong electrostatic attraction holds the sodium (Na+) and chloride (Cl-) ions in solid NaCl together. One mole of solid NaCl must dissociate into individual gaseous Na+ and Cl–ions with 769 kJ of energy (Baker et al., 1949).

$$NaCl(s) \longrightarrow Na+(g)+Cl-(g) \Delta H=769kJ$$

2.8.1.2. Cations Electronic Structures

When a main group element becomes a cation, it tends to lose all of its valence electrons, which leads to the same electrical structure as the noble gas that comes before it in the periodic table. The group numbers 2 (alkaline earth metals) and 1 (alkali metals) indicate count of valence shell electrons in their atoms and the charges of their cations upon removal.

For example, group 2 includes calcium, which has neutral atoms with 20 electrons each. At the ground state, the electronic configuration of calcium is $1s^2 2s^2 2p^6 3s^2 3p^6 4s^2$. The calcium atom becomes an eighteen-electron cation when its valence electrons are released. This cation has an electron configuration of $1s^2 2s^2 2p^6 3s^2 3p^6$ and a charge of 2+. The noble gas Ar and the Ca^{2+} ion have the same electron configuration (Ling et al., 2020).

The atoms of elements in the fourth and subsequent periods have 10 times less valence electrons than do group numbers 13–17, indicating the possibility of entire d subshells in these atoms. Consequently, the charge of a cation resulting from the complete loss of valence electrons is the group number less 10. Consider aluminum as an example, which is classified in group 13. It possesses the capacity to generate ions with a positive charge of 3, which are referred to as Al^{3+}.

Elements that are located toward the lower end of the periodic table groups exhibit behavior that occasionally deviates from what is generally predicted. Moreover, Sn^{2+}, Pb^{2+}, Tl^{+}, and Bi^{3+}ions, which are a product of a partial loss of valence shell electrons, can occur in conjunction with the presence of Sn^{4+}, Tl^{3+}, Pb^{4+}, and Bi^{5+} ions. The inert pair effect is what forms these 1+, 2+, and 3+ cations (Ling et al., 2020).

The observed phenomenon is caused by the valence s-electron pair having a comparably low energy in elements belonging to groups 13, 14, and 15 with high atomic weights. Mercury, which belongs to group 12, exhibits a remarkable trait: it may create a diatomic ion, $Hg^{22+}Hg^{22+}$ (consisting of two mercury atoms bonded together), in addition to the expected monatomic ion Hg^{2+} (produced by a single mercury atom) (Wojciechowski et al., 2003).

The behavior of inner transition and transition metal elements diverges from that of main group elements. Because they lose their outermost s electron or electrons, transition metal cations usually have charges of 2+ or 3+. They may also lose one or two d electrons from the shell that is next to the outermost. Iron, with the electron configuration $1s^2 2s^2 2p^6 3s^2 3p^6 3d^6 4s^2$, can lose its 4s electrons to produce the Fe^{2+} ion ($s^2 2s^2 2p^6 3s^2 3p^6 3d^6$). Additionally, it can lose one of its 3d electrons along with the 4s electrons to generate the Fe^{3+} ion ($1s^2 2s^2 2p^6 3s^2 3p^6 3d^5$).

It is common for the d orbitals of transition elements to be filled in the last stage of electron configuration creation. However, the outermost electrons to lose during the ionization process are the ones that are lost first. When ionized, inner transition metals usually exhibit a 3+ cationic charge. The extraction of the outermost s electrons and either *f or d* electron results in this charge (Radom et al., 1973).

2.8.1.3. Anions Electronic Structures

When nonmetal atoms have enough electrons to fill their outer p and s orbitals entirely, they approach the electron configuration of the closest noble gas, forming monatomic anions. As a result, figuring out a negative ion's charge is simple: The charge is determined by how many electrons are needed to complete the p and s orbitals of the parent atom.

Oxygen anion electron configuration is $1s^2 2s^2 2p^6$, which is also the electron configuration of neon. The oxygen electronic configuration is $1s^2 2s^2 2p^4$. The oxide ion bears the charge 2–(O^{2-}) because valence orbitals need two additional electrons (Falcetta et al., 2014).

2.8.2. Covalent Bonding

When valence electrons are shared equally between two atoms, resulting in the creation of covalent bond.

Covalency refers to the degree of electron sharing between atoms during the formation of a covalent bond. Atoms of similar or different elements can create a covalent bond. Electrons establish many forms of bonds when they are shared between two atoms. A single bond is created when two electrons are shared. A double bond is created when four electrons are shared. A triple bond is created when six electrons are shared.

Covalent Compounds characteristics:

1 These can be found in the liquid, solid, and gaseous phases of matter.
2 They are typically pliable.
3 The nature of them is non-crystalline.
4 In general, these are soluble in non-polar solvents like CCl4, benzene, ethers, toluene, gasoline, diesel, kerosene, etc., but insoluble in polar solvents like water.
5 The melting and boiling points of these are low.
6 The concentrations of them are low.
7 They are poor electrical and heat conductors.
8 These substances might exhibit isomerism.
9 In general, these are not polar (Owen and Thornley, 1966).

Ions of opposite charges are attracted to one another electrostatically and produce ions by the transfer of electrons between nonmetallic and metallic atoms. This process is known as ionic bonding. Atoms form a distinct type of bond when they are mutually attracted to a pair of electrons that they share. These kinds of bonds are referred to as covalent bonds.

Covalent bonds (Figure 2.8) form when two atoms have similar electron affinities, meaning their ionization energy and electron affinities are either equal or highly similar. For example, the union of two hydrogen atoms by covalent bonding results in the formation of an H2 molecule. The H2 molecule is able to maintain its stability by providing two electrons to each hydrogen atom, resulting in both atoms having the equal number of valence electrons as a noble gas He (Nordholm and Bacskay, 2020).

Unlike ionic molecules, covalently bound compounds have unique physical characteristics. Because the intermolecular forces between electrically neutral molecules in covalent compounds are relatively weaker than those between electrically charged ions in ionic compounds, covalent compounds typically exhibit significantly lower melting and boiling temperatures compared to ionic compounds. At normal temperature, many covalent compounds exist in the liquid or gaseous state.

When these compounds are in a solid form, they tend to be considerably less rigid compared to ionic solids. Furthermore, because of their electrical neutrality, the majority of covalent compounds are insoluble in water and therefore poor electrical conductors under any circumstances. On the other hand, when dissolved in water, ionic compounds function as effective electrical conductors (Nordholm and Bacskay, 2020).

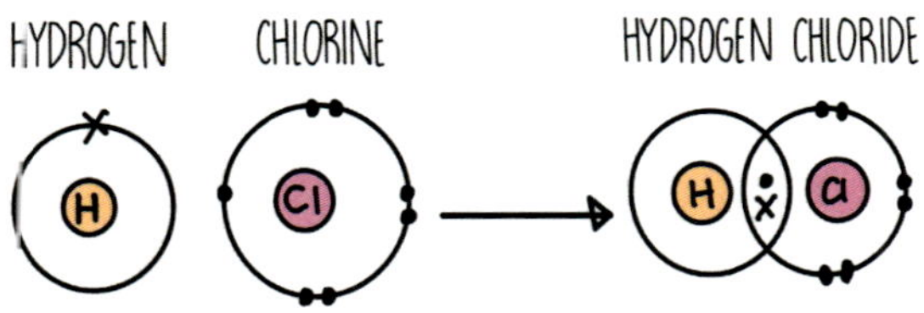

Figure 2.8. *Figure of covalent bond.*

Source: Squarespace, Creative Commons License.

2.8.2.1. Covalent Bonds Formation

Nonmetal atoms typically bond with one another in a covalent manner. A covalent bond, for example, binds the two hydrogen atoms that make up the hydrogen molecule, or H_2. This link formed for the reason depicted in Figure 2.9. On the extreme right, represented by the red line, are the two distinct hydrogen atoms, each with a unique potential energy. The x-axis shows how far apart the two atoms are from one another. Moving left down the x-axis, the two atoms approach each other, and their valence orbitals (1s) start to overlap.

The individual electrons of each hydrogen atom occupy the region surrounding the two atoms while interacting with their respective atomic nuclei. The shared electron exhibits a significant affinity towards both nuclei, resulting in system stabilization. As the bond distance reduces, the potential energy also falls. Potential energy increases when atoms get closer to one another because the positive charges in their nuclei begin to repel one another. The bond length is defined as the distance at which the minimal potential energy is obtained (Zeng et al., 2015).

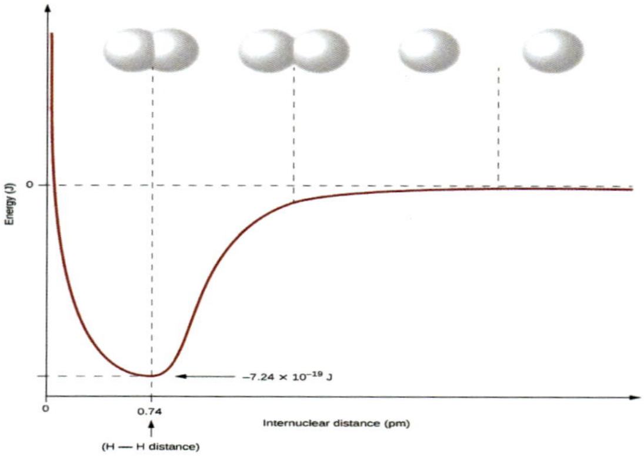

Figure 2.9. *When two independent hydrogen atoms get closer to one another, their potential energy drops and their single electrons share to create a covalent connection. The internuclear distance at which the lowest possible energy is obtained is known as the bond.*

It is important to remember that breaking chemical bonds requires energy, and this is an energy-absorbing process. Conversely, energy is released when chemical bonds are formed, leading to an energy release. The covalent bond between H_2 atoms is quite strong. For the molecules to split, the bonds between the atoms must break and 436 kJ of energy must be released by one mole of hydrogen:

$$H_2(g) \longrightarrow 2H(g) \Delta H = 436 kJ$$

Conversely, when one mole of H_2 molecules is created from two moles of H atoms, the same amount of energy is released (Tse et al., 1980):

$$2H(g) \longrightarrow H_2(g) \Delta H = -436 kJ$$

2.8.2.2. Polar versus Pure Covalent Bonds

In diatomic molecules, such as Cl_2 and H_2, the electrons are shared and distributed evenly across the atoms to form covalent bonds. This bond is commonly known as a pure covalent bond. In pure covalent bonds, shared electrons have an equal chance of being found near either nucleus.

When it comes to Cl_2, each atom has seven valence electrons to begin with. Each Cl atom shares one electron with the other to establish a covalent bond.

$$Cl + Cl \longrightarrow Cl_2 Cl + Cl \longrightarrow Cl_2$$

Eight electrons in total, two shared (bonding) electrons and six nonbonding electrons make up each atom. This sum corresponds to the number of valence electrons in argon, a noble gas. Cl_2 also displays a pure covalent bond because the bonding atoms are of the same kind.

When two separate atoms form a bond, the electrons involved in the covalent connection are shared, but not in an equal manner. The electron density is redistributed in favor of that specific atom due to the stronger attraction of the bonding electrons towards it compared to the other atoms. When there is an uneven distribution of electrons, one atom exhibits a partial positive charge while the other atom exhibits a partial negative charge. This is referred to as a polar covalent bond (Coll and Treagust, 2002).

A partial negative charge will be acquired by the atom with the higher affinity for electrons, whereas a partial positive charge will be acquired by the other atom. An atom's electronegativity measures its ability to attract electrons. It determines how atoms in a chemical bond share electron. Atoms attract electrons into bonds due to their electronegativity.

The electron distribution inside the H–Cl bond is shown in Figure 2.10. It is notable that the area shaded around H is much smaller than the area colored around Cl. Compare this with the uniform electron distribution of the nonpolar H_2 bond shown in Figure 7.4.

Using the lowercase Greek letter "delta," δ, in conjunction with a plus or minus sign, we may determine which atom is positive and which is negative in a polar covalent connection. A (δ+) plus sign is used to represent partial positive charge and a partial negative charge (δ–) is indicated by a minus sign. The H–Cl molecule's iconography is displayed in Figure 2.10 (Jonsson et al., 1994).

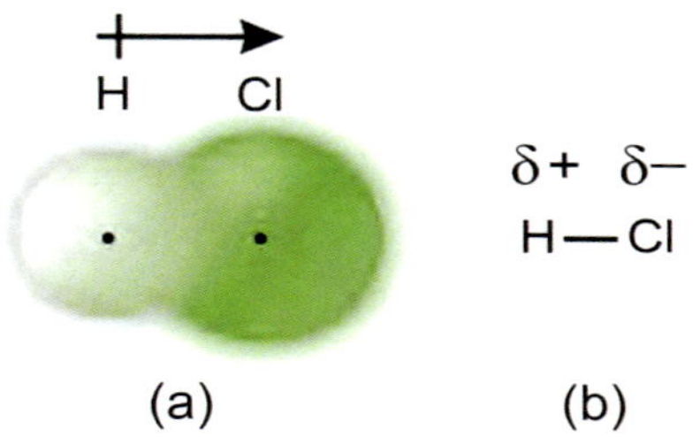

Figure 2.10. *The HCl molecule has an irregular distribution of electron density. The area surrounding the chlorine nucleus has a higher electron density. The chlorine and hydrogen nuclei location in the molecule is indicated by the tiny, black dots. (b) The symbols δ– and δ+ represent the polarity of the H–Cl bond.*

Source: Pressbooks, Creative Commons License.

2.8.2.3. Electronegativity

The polarity of a bond, whether it is polar or nonpolar covalent, is determined by the electronegativity of the atoms involved in the bonding. An atom's electronegativity measures its ability to attract electrons. It determines how atoms in a chemical bond share electrons. Atoms attract electrons into bonds due to their electronegativity.

In a polar covalent bond, electrons are transferred toward the atom with higher electronegativity. Consequently, a partial negative charge is carried by the atom with the higher electronegativity. The electron distribution becomes increasingly polarized and the atoms' partial charges grow as the electronegativity difference rises (Bergmann and Hinze,1996).

Figure 2.11 displays elemental electronegativity values as reported by 20th-century chemist Linus Pauling. Electronegativity on the periodic table typically increases with time from left to right and decreases as one proceeds down a group. Consequently, the nonmetals (upper right) typically have the largest electronegativities. Fluorine is the most electronegative element among them, having an electronegativity value of 4.0.

Metals generally have lower electronegativities, with group 1 metals exhibiting the lowest electronegativities. It should be noted that noble gases are not included in this picture since these atoms typically do not engage in electron sharing with other atoms due to their complete valence shell. (While it is true that noble gas compounds such as XeO_2 exist, their creation is restricted to highly unusual circumstances, which makes them a rare case in terms of the overall concept of electronegativity) (Hinze and Jaffe, 1962).

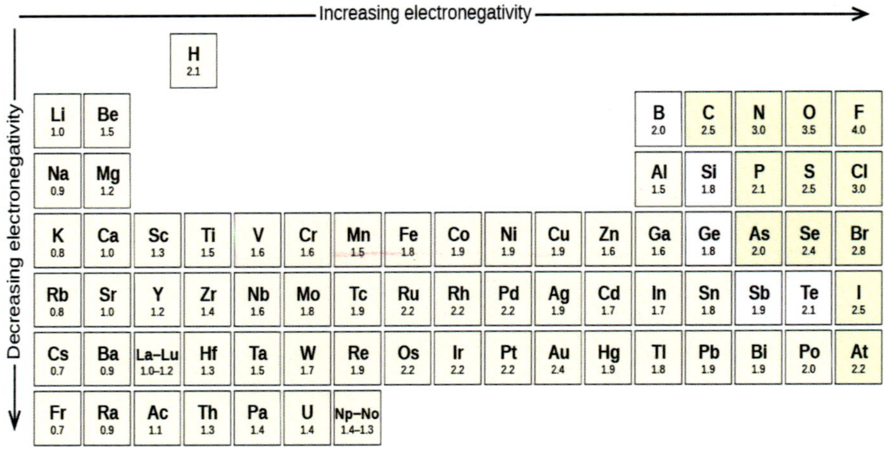

Figure 2.11. *Pauling's determination of electronegativity values exhibits consistent periodic patterns, with higher electro-negativities located in the top right quadrant of the periodic table.*

Source: Chemistry Libretexts, Creative Commons License.

2.8.2.4. Electronegativity Versus Electron Affinity

It is essential to comprehend the distinction between electron affinity and electronegativity. The electron affinity of an element is a physical quantity that is measured and expressed in kJ/mol. It stands for the energy generated or absorbed when an electron is added to an isolated gas-phase atom. Conversely, the strength with which an atom draws electrons to a bond is described by its electronegativity. It is an estimated, unmeasured, dimensionless quantity. Pauling compared the energies needed to break various kinds of bonds in order to determine the first values for electronegativity. He randomly chose a relative scale with a range of 0 to 4 (Zhan et al., 2003).

2.8.2.5. Electronegativity and Bond Type

The electronegativity difference between two bound atoms can estimate bond polarity and type. When there is a minimal or negligible difference, the bond falls under the category of covalent and nonpolar. As the bond grows larger, it can be classified as either polar covalent or ionic. The varying electronegativity differences between the atoms in different bonds lead to distinct types of bonding. There are three types of bonds: nonpolar, polar ionic and covalent.

The extent of electron sharing between atoms ranges from total equality (pure covalent bonding) to complete absence (ionic bonding). Figure 2.12 shows the correlation between the difference in electronegativity and the type of bond (Pauling, 1932).

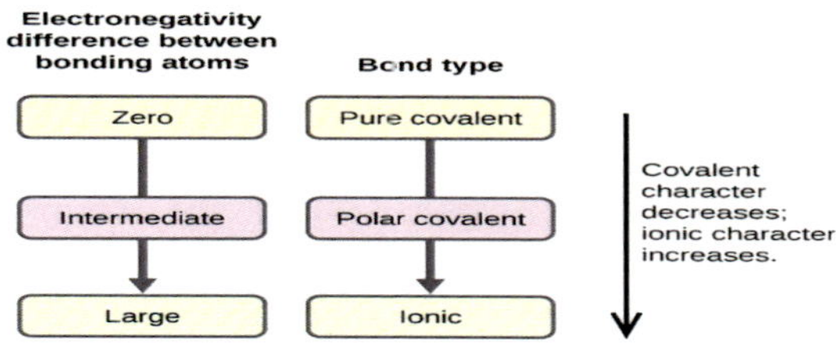

Bond Type	Electronegativity Difference
pure covalent	< 0.4
polar covalent	between 0.4 and 1.8
ionic	> 1.8

Figure 2.12. *The bond progressively becomes ionic as the difference between the electronegativity of two atoms increases.*

Source: RCS Education , Creative Commons License.

Figure 2.12 shows how polar covalent and ionic bonds vary in electronegativity. This table is a general reference, with a few exceptions. The N and H atoms in NH_3 and H and F in HF. Although the electronegativity variances between these compounds are 0.9 and 1.9, they both form polar covalent bonds. Both MnI_2 and NaCl form ionic compounds, despite the electronegativity difference between the Na and Cl atoms and the Mn and I atoms being 2.1 and 1.0, respectively.

In order to determine whether a bond is ionic or covalent, it is necessary to assess the types of atoms involved in the bond and their respective positions on the periodic table. Covalent bonds typically form between two nonmetals, while ionic bonds commonly occur between a nonmetal and metal. Some compounds exhibit both ionic and covalent bonding. Polar covalent bonds are responsible for the attraction between atoms in polyatomic ions like NO_3^-, $OH^-NO_3^-$, and NH_4^+, together. But when these polyatomic ions mix with ions that have the opposite charge, they create ionic compounds.

For example, potassium nitrate, represented by the chemical formula KNO_3, is composed of the potassium cation (K^+) and the polyatomic nitrate anion (NO_3^-). The electrostatic interaction between the K^+ and NO_3^- ions in potassium nitrate leads to mostly ionic bonding. In addition, the nitrogen and oxygen atoms in the NO_3– ion display a covalent connection (Sproul, 2001).

Example: I: Formation of H_2 molecule. The electronic configuration of 'H' is

$_1H - 1s^1$

The electronic structure of the element 'H' shows that it has 1 valence electron and needs one more electron to achieve a duplet form. Consequently, every hydrogen atom mutually shares its electron to establish a covalent link

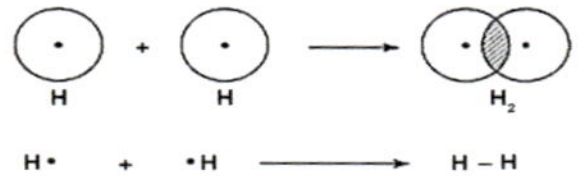

Example: 2: Cl_2 molecule Formation. The 'Cl' electronic configuration is

$_{17}Cl - 1s^2 2s^2 2p^6 3s^2 3p^5$

The electron configuration of 'Cl' indicates that it has 7 valence electrons and need one more electron to attain an octet. As a result, each chlorine atom shares one of its valence electrons with every other chlorine atom, leading to the creation of a covalent bond (Figure 2.4) (Smith, 2007).

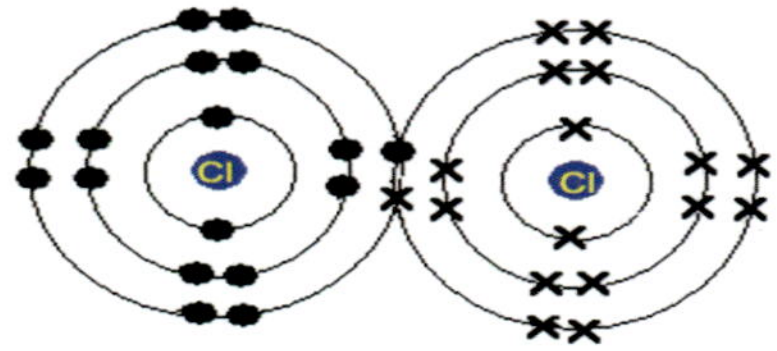

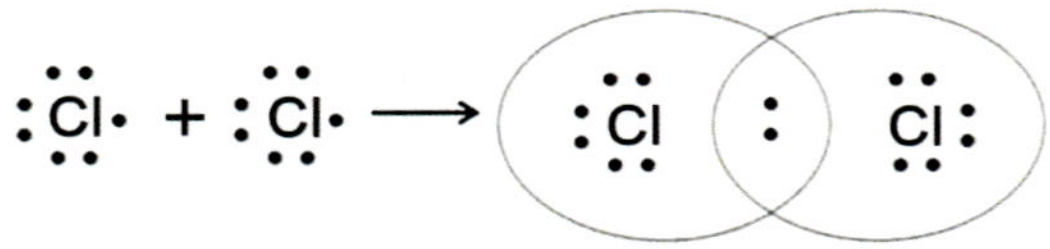

Example:3: Process of O_2 molecule formation. The electronic configuration of 'O' is

$_8O - 1s^2 2s^2 2p^4$

The electrical structure of 'O' indicates the presence of 6 valence electrons and necessitates the addition of two additional electrons to achieve an octet. Consequently, every oxygen atom mutually shares two of its valence electrons, resulting in the formation of a double covalent bond (Figure 2.5) (Meek and Garner, 2005).

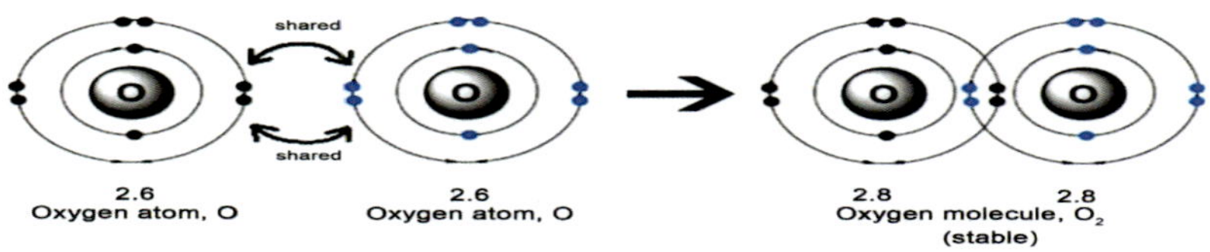

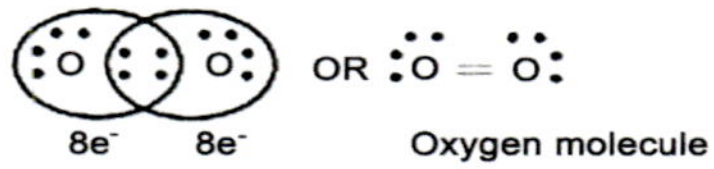

2.8.3. Co-Ordinate Bond

A co-ordinate or dative bond is a chemical link created when two atoms (or ions) partially donate and partially share a single pair of electrons.

Criteria for the formation of a coordinate or dative bond:

There should be at least one lone or unshared pair of electrons in one of the participating atoms. The distance between the other atom and the closest inert gas element should be less than two electrons. Both combining atoms share a portion of the single electron pair on one atom. An arrow (→) designating a coordinate bond indicates the acceptor atom by its head and the donor atom by its tail.

Due to its partial polarity, the co-ordinate bond is alternatively referred to as a dative or semi-polar bond, or a co-ionic bond (Harrowfield,1996).

Example: 1: Ammonium ion ($NH_4{}^+$) Formation.

The ammonium ion (NH^{4+}) is created through the union of the NH and H^+ ions. The reaction between NH_3 and H^+ results in the formation of $NH_4{}^+$.

While the H+' ion has no electrons and needs two to become duplet, ammonia (NH_3) has a single pair of electrons over N.' As a result, a coordinate bond is created when the unshared pair of electrons over nitrogen in NH_3 is partially shared with the H^+ ion.

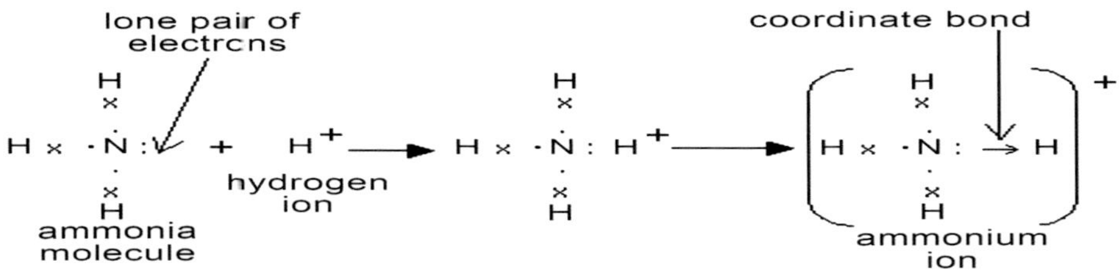

Example: 2: Sulphur dioxide (SO_2) molecule Formation (Bond, 1966).

The electronic configurations of O and S are as follows: for 'O,' it is $1s^2 2s^2 2p^4$ and '$_6$S,' it is $1s^2 2s^2 2p^6 3s^2 3p^4$.

Six valence electrons are present in each of S' and O,' according to the electrical configurations. By sharing two of its valence electrons with two of the oxygen atom's valence electrons, the central sulfur atom creates a double covalent bond. To create a co-ordinate connection, one of the central sulfur atom's lone pair of electrons is partially shared with another oxygen atom (Figure 1.12)

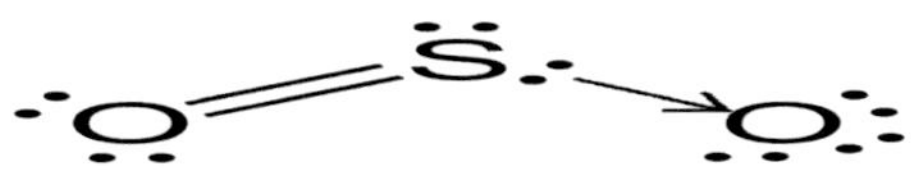

The features of a Coordinate Compound are as follows:

1 These connections exhibit both rigidity and directionality.

2 Coordinate bonds do not undergo ionization in a fused state or solution.

3 Typically, they are not able to dissolve in water but can dissolve in solvents that are not polar.

4 The boiling and melting temperatures of metallic compounds are intermediate between those of covalent compounds and ionic compounds.

5 These compounds are classified as semi-polar, meaning they have a higher level of polarity compared to covalent compounds but a lower level of polarity compared to ionic compounds.

6 They exhibit isomerism (Ahrland and Chatt, 1957).

2.8.4. Lewis Symbols and Structure

Following section explains Lewis Symbols and Structure.

2.8.4.1. Lewis Symbols

The electrons in the valence shell of a molecule can be seen clearly in a Lewis Structure. It is used to display spatial arrangement of the electrons that surround particular atoms within molecules. In the context of a chemical bond, electrons are depicted as "dots" or as a line that joins the two atoms. For examples:

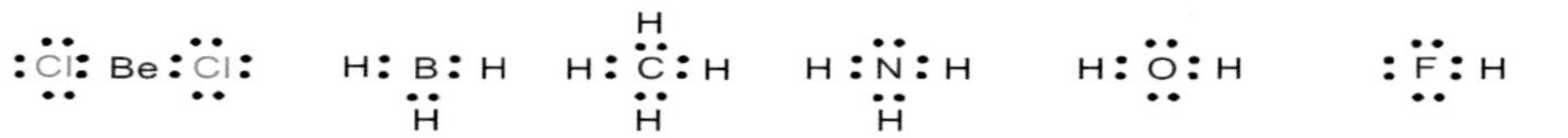

Beryllium dichloride — Borane — Methane — Ammonia — Water — Hydrofluoric acid

We have covered the several kinds of bonds that can form between atoms or ions so far in this chapter. The atoms in these bonds always swap electrons from their valence shells. The Lewis symbols and Lewis structures that are frequently used to depict valence shell electrons and chemical bonds will be covered in this section.

Atomic and monatomic ion valence electron configurations are represented by Lewis symbols. An elemental symbol and a dot for each of its valence electrons surround the elemental sign to form a Lewis symbol (Ahma and Zakaria, 2000):

$$\cdot Ca \cdot$$

Lewis symbols, like this one for calcium and sodium, can also be used to explain how cations are formed from atoms:

$$\underset{\text{sodium atom}}{Na\cdot} \longrightarrow \underset{\text{sodium cation}}{Na^+} + e^- \qquad \underset{\text{calcium atom}}{\cdot Ca \cdot} \longrightarrow \underset{\text{calcium cation}}{Ca^{2+}} + 2e^-$$

They can also be used for demonstrating how atoms generate anions, as demonstrated here for sulfur and chlorine (Ford, 1966):

Lewis symbols are used to show the transport of electrons during the creation of ionic compounds in Figure 2.13.

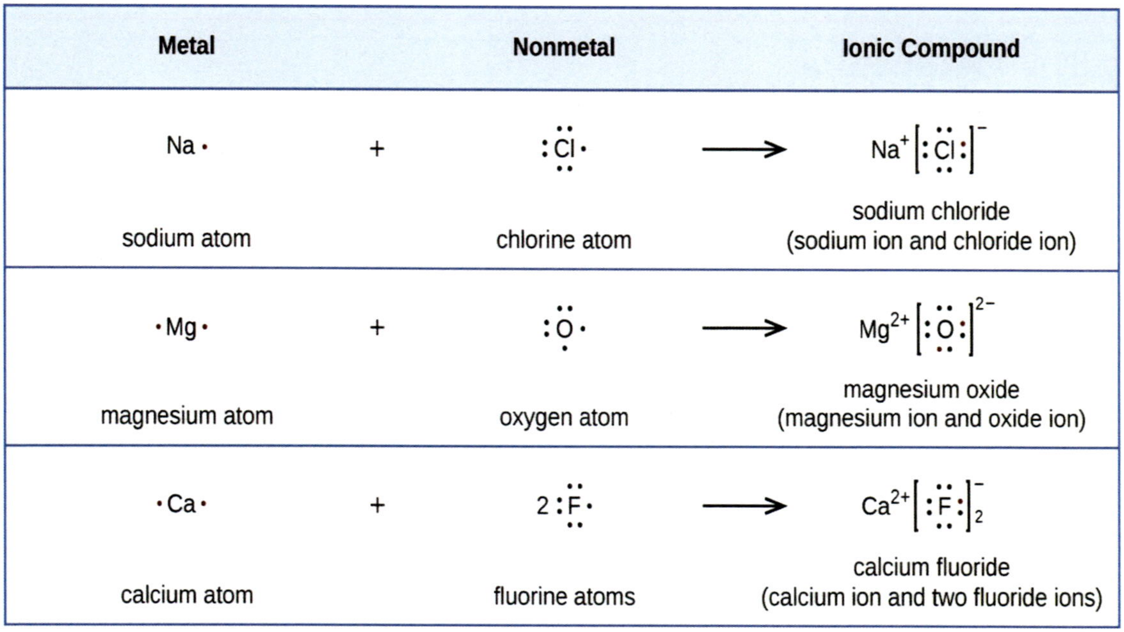

Figure 2.13. *When atoms lose electrons, cations are formed; less Lewis dots indicate this, whereas anions are formed when there are more Lewis dots. The overall number of electrons remains constant (Horton, 1939).*

2.8.4.2. Lewis Structures

Lewis symbols are essential in representing molecular and polyatomic ion bonding through visual representations known as Lewis structures. They can also be used to signal the creation of covalent connections. Two chlorine atoms share one pair of electrons when they unite to form a chlorine molecule, which is an example of electron sharing (Cooper et al., 2012).

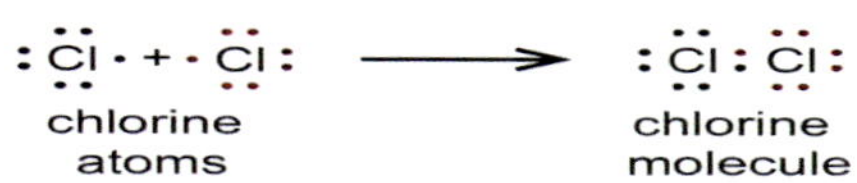

According to the Lewis structure, each chlorine atom contains three unshared electron pairs, or "lone pairs," and one shared electron pair, which is represented as a line connecting the atoms. In certain cases, a hyphen (or horizontal line) is used to symbolize a pair of electrons that are shared by both parties:

Two atoms sharing two electrons make form a single bond. The two electrons in a single bond and the six electrons in lone pairs make up the eight valence electrons that interact with each chlorine (Cl) atom (Finkenstaedt-Quinn et al., 2019).

2.8.5. The Octet Rule

Similar to the chlorine molecule, the other halogen molecules (Br_2, I_2, At_2 and F_2) form bonds with three lone pairs of electrons per atom and one single connection between atoms. Thus, every halogen atom can have the electron configuration of a noble gas. According to the octet rule, atoms in the main group will always seek to form bonds in order to acquire eight valence electrons in total.

Typically, the determination of the number of bonds an atom can form relies on the number of electrons needed to reach an octet, which is comprised of eight valence electrons. The principle discussed here is especially relevant to the nonmetals located in the second period of the periodic table, including carbon (C), nitrogen (N), oxygen (O), and fluorine (F). For example, every atom of a group 14 element possesses four electrons in its outermost shell, necessitating an additional four electrons to achieve an octet.

As demonstrated by the examples of silicon in SiH_4 (silane) and carbon in CCl_4 (carbon tetrachloride), four electrons can be obtained by creating four covalent bonds. Hydrogen, unlike other elements, does not follow the octet rule since it only requires two electrons to complete its outermost shell. Transition elements and inner transition elements violate the octet rule (Quayle et al., 2015).

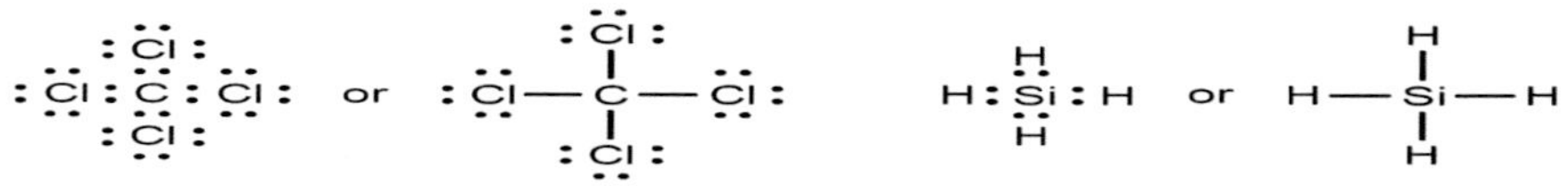

carbon tetrachloride silane

The atomic Lewis symbol of Group 15 elements, including nitrogen, consists of five valence electrons. Three unpaired electrons and one lone pair are among them. These atoms form three covalent bonds in order to gain an octet, as shown by NH_3 (ammonia). All of the atoms in group 16, including oxygen, establish two covalent connections to form an octet configuration (Gillespie and Robinson, 2007).

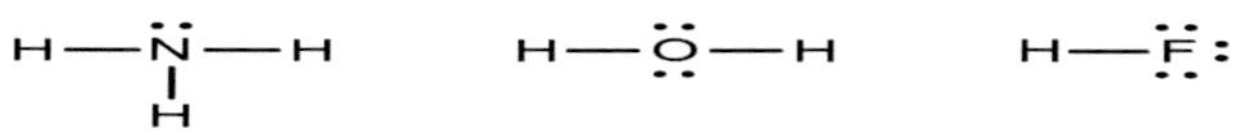

ammonia Water hydrogen fluoride

2.8.6. Double and Triple Bonds

As previously mentioned, a single bond is created when two atoms share a single pair of electrons. However, in order for a pair of atoms to reach the required octet, it is possible that many pairs of electrons will need to be shared. Double bonds are created when two atoms share two pairs of electrons, such as the two carbon atoms in ethylene (C_2H_4) and the oxygen and carbon atoms in CH_2O (formaldehyde).

formaldehyde

ethylene

A triple bond is created when two atoms share three electron pairs, as seen in cyanide ion (CN–) and carbon monoxide (CO) (Uce, 2015).

carbon monoxide

cyanide ion

2.8.7. The Octet Rules' Exceptions

In several covalent compounds, the Lewis structures show core atoms without eight electrons. Classifying these molecules into three categories is possible:

1. Odd-electron molecules contain an unpaired electron due to their odd number of valence electrons.
2. Electron-deficient molecules are characterized by a core atom that possesses less electrons than required to achieve a noble gas structure.
3. Hypervalent molecules are characterized by a core atom that possesses an excess of electrons beyond what is required for achieving a noble gas configuration (Auvert, 2014).

2.8.7.1. Odd-electron Molecules

Molecules with an odd number of electrons are called free radicals. One example of an odd-electron molecule, also known as an unpaired electron molecule, is nitric oxide (NO). In internal combustion engines, nitrogen, and oxygen combine at high temperatures to form nitrogen oxide.

The Lewis structure of an odd-electron molecule, like NO, can be constructed using the same five steps that apply to other compounds, with a few minor adjustments:

i. **Calculate the total number of electrons in the valence (outer) shell.** The total number of valence electrons is the sum of the valence electrons from nitrogen and oxygen. Given the presence of an odd number, it can be concluded that a free radical is immediately present. This suggests that not all atoms are capable of holding eight electrons in their valence shell.

ii. **Draw the molecule's skeletal structure.** With a N–O single bond, we can quickly sketch a skeleton: N–O

iii. **On the terminal atoms, distribute the remaining electrons as lone pairs.** In this scenario, the absence of a core atom necessitates the dispersion of electrons between both atoms. In these scenarios, we assign eight electrons to the atom with higher electronegativity, resulting in oxygen having a complete valence shell:

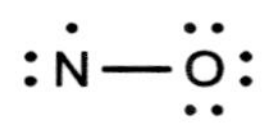

iv. **Distribute the electrons remaining to central atom.** Since there are no remaining electron, this step is not applicable.

v. **Reconfigure the electrons to form numerous connections with the center atom to achieve octets whenever possible.** We know an odd-electron molecule cannot have an octet for each atom, but we aim to get as close as possible. Five electrons surround nitrogen in this scenario. We build a NO double bond with nitrogen using one of oxygen's lone pairs to complete the nitrogen octet. Nitrogen exceeds its nine-electron capacity; hence we cannot build a triple bond with another lone pair of electrons on oxygen (Parkin, 2023).

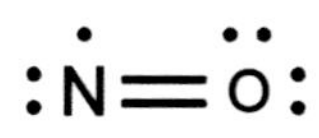

2.8.7.2. Electron-Deficient Molecules

Additionally, a few molecules with unfilled valence shells in their centers will come across our path. These are compounds that typically have hydrogen as an exterior atom, center atoms from groups 2 and 13, or other atoms that do not make multiple bonds. For example, core atoms in molecules of boron trifluoride (BF_3) and beryllium dihydride (BeH_2) have only two and three valence electrons, respectively.

The Lewis structure of BeH_2 is limited to one alternative with just H as exterior atoms, and it does not meet the octet rule for the core Be atom (see as follows). The octet rule can be satisfied by drawing a structure in which a boron and a fluorine atom form a double bond in BF_3, however experimental data suggests that the bond lengths are more in line with what would be predicted for B–F single bonds (Mahmood et al., 2021).

This indicates that a boron with an electron deficiency and three B–F single bonds make up the optimal Lewis structure. The compound's reactivity is also consistent with a boron that lacks electrons. The B-F bonds, however, are a little bit shorter than one would expect for B-F single bonds, suggesting that the molecule in question contains some double bond character.

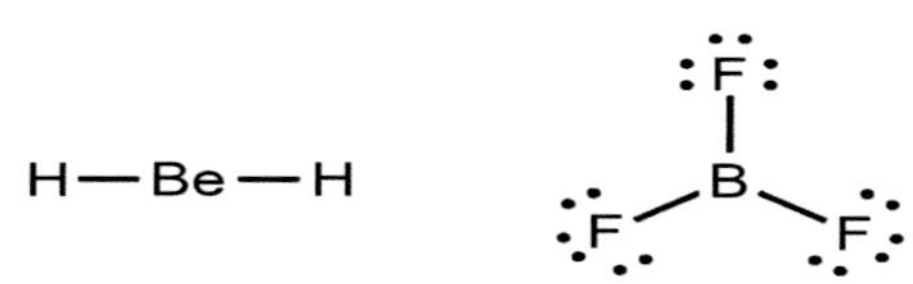

An atom, such as the boron atom in BF_3, that does not possess eight electrons, has high reactivity. It quickly forms a chemical bond with a molecule that has an atom with unshared electrons. An example of a reaction takes place between NH_3 and BF_3 as a result of the nitrogen's lone pair being capable of being shared with the boron atom (Lipscomb, 1966).

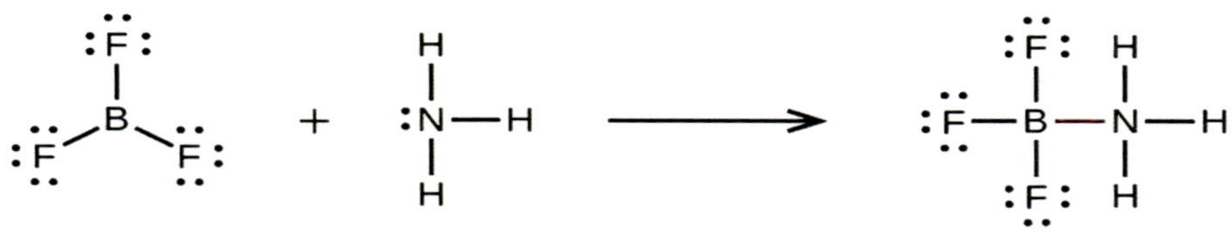

2.8.7.3. Hypervalent Molecules

The elements in the second period (n = 2) of the periodic table have valence shells that may hold up to eight electrons in their orbitals. This is due to the fact that they have four valence orbitals: three 2p orbitals and one 2s orbital. Elements in the third and subsequent periods (n > 3) have more valence orbitals and are able to produce more than four pairs of electrons with other atoms. This is as a result of their shared energy level empty d orbitals. Hypervalent molecules are a term used to describe some molecules composed of certain elements. (Reed and Schleyer, 1990) (Figure 2.14).

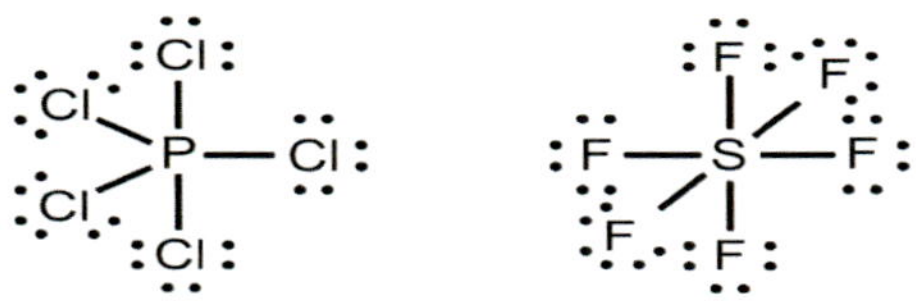

Figure 2.14. *The phosphorus atom at the center of PCl5 shares five electron pairs. Sulfur and six electron pairs are shared in SF6.*

Some hypervalent compounds, like IF_5 and XeF_4, have lone pairs of electrons in the outer shell of the central atom.

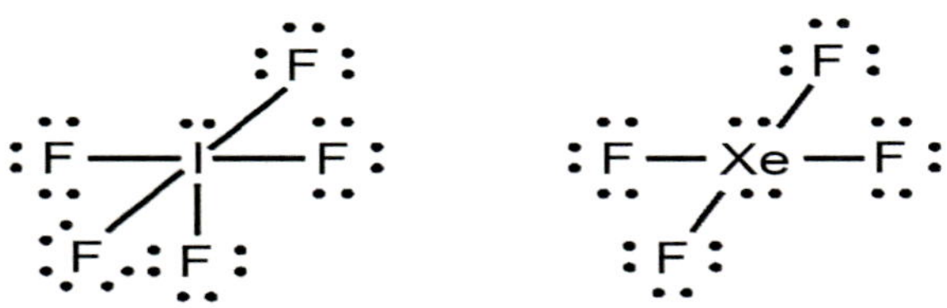

Upon further examination of the Lewis structures for these compounds, it becomes apparent that even after the addition of eight electrons to the valence shells of the outer atoms, there are still remaining electrons. The core atom must be allocated more electrons (Dobado et al., 1998).

2.8.8. Bond Strength: Covalent Bonds

The bond strength is a measure of the degree of attachment between two atoms and, consequently, the amount of energy needed to separate them. This section examines the relative strength of ionic and covalent bonds, specifically in regards to the lattice energy of a molecule.

Covalent bonding is necessary to guarantee molecules' stability. The bond dissociation energy of a covalent bond, or the energy required to break the paired atoms apart, determines the bond's strength. Energy must be added in order to attempt to dissociate a pair of chemically bonded atoms (refer to Figure 7.4). One way to determine the strength of a bond is by evaluating the energy required to break it (Grandbois et al., 1999).

Breaking a covalent bond in one mole of gaseous molecules requires bond energy. For a diatomic molecule, the bond energy, (D_{X-Y}), is the typical enthalpy change during an endothermic reaction:

$$XY(g) \rightarrow X(g) + Y(g)$$

$$D_{X-Y} = \ddot{A}H^{\circ}$$

For example, the pure covalent H–H bond, or (D_{H-H}), has a bond energy of 436 kJ per mole of broken H–H bonds:

$$H_2(g) \rightarrow 2H(g)$$

$$D_{H-H} = \ddot{A}H^{\circ} = 436 \text{ kJ}$$

Chemical bonds are present in molecules with three or more atoms at least. The standard change in enthalpy for the exothermic process that dissociates every bond in a molecule is equal to the entire bond energy in that molecule. For example, standard enthalpy change of process is equivalent to the overall energy of 1660 kJ required to dissociate the four C–H bonds in CH4 (Beyer, 2000):

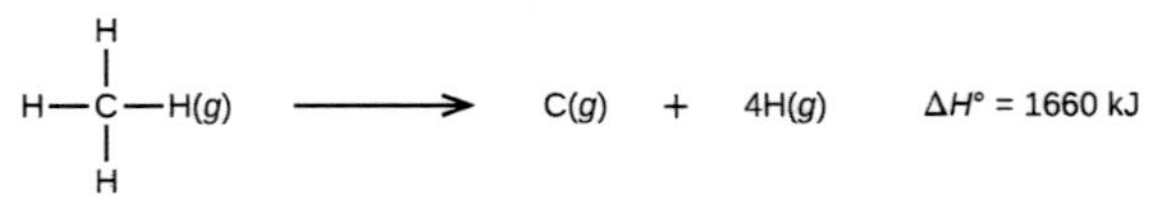

The average energy needed to break a C–H bond is 415 kJ/mol, according to the calculation. It is crucial to take into consideration for the breaking of four C-H bonds for every mole of the reaction when calculating the value. Although the molecules are identical, the original molecule with the four C-H bonds exhibits varying energy requirements for breaking them. Once the initial bond is disrupted, requiring an energy of 439 kJ/mol, the subsequent bonds become less resistant to breaking. The value of 415 kJ/mol is an approximate estimation, rather than a precise measurement, of the energy required to break a particular bond (Drago and Wong, 1995).

The bond strength between two atoms is determined by the number of electron pairs involved in the bond. Bond strength and length generally have an inverse relationship, which means that as a connection's strength increases, so does its length. As a result, triple bonds between the same pair of atoms are reported to be shorter in length and stronger than double bonds. Comparably, between the same pair of atoms, double bonds are shorter in length and stronger as compared to single bonds (Bonn and Hunger, 2021).

Table 2.1 shows the average bond energy for some common bonds, and Table 2.2 compares their lengths and strengths. The binding strength usually reduces as we proceed down the group when one atom connects to multiple atoms in the group. For example, the values of C–F, C–Cl, and C–Br are 439, 330, and 275 kJ/mol, respectively (Gavezzotti, 2016).

Table 2.1. *Tabular data of Bond Energies (kJ/mol)*

Source: Gavezzotti, Creative Commons License.

	Bond Energies (kJ/mol)				
Bond	**Bond Energy**	**Bond**	**Bond Energy**	**Bond**	**Bond Energy**
H–N	390	Si–P	215	C–I	240
H–C	415	F–Br	235	C–Cl	330
H–O	464	Si–Si	230	C–Br	275
H–H	436	Si-S	225	N-N	160
H–Si	395	Si–Cl	359	N=NN=N	418
H–F	569	F-Cl	255	C-S	260
H–Br	370	P–S	230	N–P	210
H–P	320	Si–Br	290	N≡NN≡N	946
H–S	340	Si–I	215	N–O	200
H–Cl	432	P–P	215	N–F	270
C–C	345	P–Br	270	N–Br	245
H–I	295	P–Cl	330	N–Cl	200
C=CC=C	611	P–I	215	O–O	140
C=NC=N	615	S–Br	215	O–Si	370
C≡CC≡C	837	S–S	215	O=OO=O	498
C–N	290	S–Cl	250	O–F	160
C–O	350	Cl–Br	220	O–Cl	205
C≡NC≡N	891	Cl–Cl	243	O–P	350
C≡OC≡O	1080	Br–Br	190	F–F	160
C=OC=O	741	Cl–I	210	O–I	200
C–Si	360	I–I	150	F–P	489
C–F	439	Br–I	180	F–Si	540
C–P	265			F–S	285

Table 2.2. *Tabular Data of Average Bond Energies and Bond Length for Some Common Bonds Source: Gavezzotti, Creative Commons License.*

Bond	Bond Energy (kJ/mol)	Bond Length (Å)
C–C	345	1.54
C–N	290	1.43
C≡CC≡C	837	1.20
C=NC=N	615	1.38
C–N	290	1.43

C–O	350	1.43
C≡NC≡N	891	1.16
C=OC=O	741	1.23
C≡OC≡O	1080	1.13

In situations when enthalpies of formation are not available, bond energies can be used to approximate enthalpy changes in processes. The exothermic and endothermic nature of a reaction can also be determined by these calculations. An exothermic reaction occurs when the products have stronger bonds and it requires more energy to break the bonds in the reactants. A positive change in enthalpy (ΔH) and heat absorption are characteristic features of an endothermic process. This happens when there are weaker chemical bonds between the product and reactant (Iozzi et al., 2009).

Calculating change of enthalpy, or ΔH, in a chemical reaction involves considering the energy needed to break the reactants bonds (energy "in," positive sign) and the energy released when the products bond is formed (energy "out," negative sign). The following is a quantitative representation of this:

$$\ddot{A}H = \sum D_{\text{bonds broken}} - \sum D_{\text{bonds formed}}$$

In this phrase, the symbol Σ denotes the mathematical operation of summation, which involves adding a series of numbers together. The bond energy, denoted by D, is always expressed in kilojoules per mole and is consistently positive. The bond energy can be determined by referring to a table, like Table 2.2, and will differ based on the type of bond – whether it is a triple, double or single bond (Brandhorst and Grunenberg, 2008).

When determining enthalpies using this approach, it is crucial to take into account the bonding that exists in both the reactants and products. Given that D values represent average values for a certain bond over various molecules, this approach offers an approximate estimation rather than a precise figure for the enthalpy of reaction.

Let's consider the reaction (Nordholm and Bacskay, 2020):

$$\text{H–H(g)} + \text{Cl–Cl(g)} \rightarrow 2\text{H–Cl(g)}$$

Or more conventionally written as:

$$H_2(g) + Cl_2(g) \rightarrow 2HCl(g)$$

In this reaction, one mole of hydrogen molecules (H_2) reacts with one mole of chlorine molecules (Cl_2) to form two moles of hydrogen chloride (HCl). The bonds broken in this reaction are one H–H bond and one Cl–Cl bond, while the bonds formed are two H–Cl bonds.

To determine the enthalpy change (ΔH) for this reaction, you would use the formula:

$$\Delta H = \left(D_{H-H} + D_{Cl-Cl}\right) - 2 \times D_{H-Cl}$$

Two moles of hydrochloric acid (HCl) necessitate the breaking of one mole of hydrogen-hydrogen (H-H) bonds and one mole of chlorine-chlorine (Cl-Cl) bonds. Breaking these chemical bonds requires the same energy as the Cl–Cl and H–H bonds (243 and 436 kJ/mol, respectively). Two moles of H–Cl bonds with 432 kJ/mol energies are formed. An overall energy discharge of 864 kJ is the consequence of this. The reaction leads to a net release of energy as a result of the enhanced strength of bonds in products in comparison to bonds in reactants:

Here is the equation rewritten as requested:

$$\Delta H = \sum D_{bonds\ broken} - \sum D_{bonds\ formed}$$

$$\Delta H = \left[\Delta H_{H-H} + \Delta H_{Cl-Cl}\right] - 2\Delta H_{H-Cl}$$

$$\Delta H = \left[436 + 243\right] - 2\left(432\right) = -185\,kJ$$

The release of excess energy as heat indicates that the reaction is exothermic (Nordholm and Bacskay, 2020).

ACTIVITY 2.1.

Explore the Aufbau principle by constructing electron configurations for different atoms. Use periodic tables and electron configuration rules to predict how electrons fill atomic orbitals, influencing chemical reactivity and bonding.

SUMMARY

- Chapter Two of "Atomic Structure and Bonding" delves into foundational concepts crucial to understanding the composition and behavior of atoms. It begins with an exploration of atomic structure, detailing the roles of protons, neutrons, and electrons. Rutherford's landmark Gold-foil experiment is examined, highlighting its contributions to atomic theory and the subsequent formulation of Rutherford's atomic model, albeit with acknowledged limitations.
- Bohr's atomic model is then introduced, elucidating its postulates and its role in refining our understanding of atomic structure. The chapter proceeds to clarify key distinctions between atomic number (Z) and mass number (A), essential for identifying and categorizing elements. Additionally, the concept of isotopes, along with related terms like isobars and isotones, is discussed to underscore their significance in isotopic diversity.
- The Aufbau principle and its application in determining electron configurations are detailed, setting the stage for a comprehensive examination of chemical bonding. Various types of bonding mechanisms are explored, including ionic, covalent, coordinate, and their respective properties such as bond strength and exceptions to the octet rule.

REVIEW QUESTIONS

1. Explain Rutherford's Gold-foil experiment and its significance in the development of atomic theory.
2. Compare and contrast Rutherford's atomic model with Bohr's atomic model.
3. Define atomic number (Z) and mass number (A). How are they different from each other?
4. Discuss the concept of isotopes. Provide examples and explain their relevance.
5. How does the Aufbau principle help in determining electron configurations? Provide an example.
6. Describe the types of chemical bonding discussed in the chapter. Include their characteristics and examples.
7. What are the exceptions to the octet rule in covalent bonding? Explain with examples.

MULTIPLE CHOICE QUESTIONS

1. Which experiment led to the discovery of the atomic nucleus?

a. Millikan oil drop experiment

b. Rutherford's Gold-foil experiment

c. Bohr's atomic model

d. Thomson's cathode ray experiment

2. **Which scientist proposed a planetary model of the atom with electrons orbiting the nucleus in fixed paths?**
   a. Rutherford
   b. Bohr
   c. Thomson
   d. Dalton
3. **What does the atomic number of an element represent?**
   a. Number of protons
   b. Number of neutrons
   c. Number of electrons
   d. Total number of nucleons
4. **Which term describes atoms of the same element with different numbers of neutrons?**
   a. Isotopes
   b. Isobars
   c. Isomers
   d. Isotones
5. **According to the Aufbau principle, which orbital is filled first?**
   a. p
   b. d
   c. s
   d. f
6. **Which type of bonding involves the transfer of electrons from one atom to another?**
   a. Covalent bonding
   b. Metallic bonding
   c. Ionic bonding
   d. Hydrogen bonding
7. **Which of the following is an exception to the octet rule?**
   a. CO_2
   b. H_2O
   c. BF_3
   d. CH_4

Answers to Multiple Questions

1. (b) 2. (b) 3. (a) 4. (a) 5. (c)
6. (c) 7. (c)

REFERENCES

1. Abdoolatiff, S., & Narod, F. B. (2009). Investigating the effectiveness of computer simulations in the teaching of "Atomic Structure and Bonding." *Chemistry Education in the ICT Age, 9*, 85–100.
2. Ahmad, W. Y., & Zakaria, M. B. (2000). Drawing Lewis structures from Lewis symbols: A direct electron-pairing approach. *Journal of Chemical Education, 77*(3), 329.
3. Ahrland, S., & Chatt, J. (1957). The nature of the co-ordinate link. Part X. The effect of ligands on the redox potentials of non-ionic platinous complexes and the evidence it gives for the effect of dative π-bonding on the total transfer of charge in co-ordinate bonds. *Journal of the Chemical Society (Resumed)*, 1379–1386.
4. Aman, I. K. (2020). Changes occur from mixing chemical compounds: Electrovalent bonds and covalent bonds. *International Journal Papier Advance and Scientific Review, 1*(1), 8–13.
5. Auvert, G. (2014). Improvement of the Lewis-Abegg-Octet Rule using an "even-odd" rule in chemical structural formulas: Application to hypo and hyper-valences of stable uncharged gaseous single-bonded molecules with main group elements. *Open Journal of Physical Chemistry, 4*(02), 60.
6. Bacskay, G. B., & Nordholm, S. (2013). Covalent bonding: The fundamental role of the kinetic energy. *The Journal of Physical Chemistry A, 117*(33), 7946–7958.
7. Baker, W., Ollis, W. D., & Poole, V. D. (1949). Cyclic meso-ionic compounds. Part I. The structure of the sydnones and related compounds. *Journal of the Chemical Society (Resumed)*, 307–314.
8. Bergmann, D., & Hinze, J. (1996). Electronegativity and molecular properties. *Angewandte Chemie International Edition in English, 35*(2), 150–163.
9. Beyer, M. K. (2000). The mechanical strength of a covalent bond calculated by density functional theory. *The Journal of Chemical Physics, 112*(17), 7307–7312.
10. Bidlingmeyer, B. A. (1980). Separation of ionic compounds by reversed-phase liquid chromatography: An update of ion-pairing techniques. *Journal of Chromatographic Science, 18*(10), 525–539.
11. Bohr, N. (1915). On the quantum theory of radiation and the structure of the atom. *The London, Edinburgh, and Dublin Philosophical Magazine and Journal of Science, 30*(177), 394–415.
12. Bond, G. C. (1966). Adsorption and co-ordination of unsaturated hydrocarbons with metal surfaces and metal atoms. *Discussions of the Faraday Society, 41*, 200–214.
13. Bonn, M., & Hunger, J. (2021). Between a hydrogen and a covalent bond. *Science, 371*(6525), 123–124.
14. Brandhorst, K., & Grunenberg, J. (2008). How strong is it? The interpretation of force and compliance constants as bond strength descriptors. *Chemical Society Reviews, 37*(8), 1558–1567.
15. Brik, M. G., & Srivastava, A. M. (2017). Critical review—A review of the electronic structure and optical properties of ions with d^3 electron configuration (V^{2+}, Cr^{3+},

Mn^{4+}, Fe^{5+}) and main related misconceptions. *ECS Journal of Solid State Science and Technology, 7*(1), 3079.

16. Buist, M., Bernard, S. Nguyen, T. V., Moore, G., & Anderson, J. (2004). Association between clinically abnormal observations and subsequent in-hospital mortality: A prospective study. *Resuscitation, 62*(2), 137–141.
17. Chiang, W. T., Yang, S. N., Tiator, L., & Drechsel, D. (2002). An isobar model for ⊠ photo-and electroproduction on the nucleon. *Nuclear Physics A, 700*(1, 2), 429–453.
18. Choi, Y., Hahn, C., Yoon, J. W., & Song, S. H. (2018). Observation of an anti-PT-symmetric exceptional point and energy-difference conserving dynamics in electrical circuit resonators. *Nature Communications, 9*(1), 2182.
19. Coll, R. K., & Treagust, D. F. (2002). Exploring tertiary students' understanding of covalent bonding. *Research in Science & Technological Education, 20*(2), 241–267.
20. Cooper, M. M., Underwood, S. M., & Hilley, C. Z. (2012). Development and validation of the implicit information from Lewis structures instrument (IILSI): Do students connect structures with properties? *Chemistry Education Research and Practice, 13*(3), 195–200.
21. Darrigol, O. (2013). Classical concepts in Bohr's atomic theory (1913–1925). In *Quantum Mechanics* (Vol. 2, pp. 39–61). Routledge.
22. Dawson, T. E., Mambelli, S., Plamboeck, A. H., Templer, P. H., & Tu, K. P. (2002). Stable isotopes in plant ecology. *Annual Review of Ecology and Systematics, 33*(1), 507–559.
23. Dobado, J. A., Martínez-García, H., Molina, & Sundberg, M. R. (1998). Chemical bonding in hypervalent molecules revised: Application of the atoms in molecules theory to Y_3X and Y_3XZ (Y = H or CH_3; X = N, P or As; Z = O or S) compounds. *Journal of the American Chemical Society, 120*(33), 8461–8471.
24. Drago, R. S., & Wong, N. M. (1995). Electrostatic, covalent, and transfer interpretation of bond strengths. *Inorganic Chemistry, 34*(15), 4004–4007.
25. Falcetta, M. F., DiFalco, L. A., Ackerman, D. S., Barlow, J. C., & Jordan, K. D. (2014). Assessment of various electronic structure methods for characterizing temporary anion states: Application to the ground state anions of N_2, C_2H_2, C_2H_4, and C_6H_6. *The Journal of Physical Chemistry A, 118*(35), 7489–7497.
26. Finkenstaedt-Quinn, S. A., Snyder-White, E. P., Connor, M. C., Gere, A. R., & Shultz, G. V. (2019). Characterizing peer review comments and revision from a writing-to-learn assignment focused on Lewis structures. *Journal of Chemical Education, 96*(2), 227–237.
27. Ford, L. S. (1966). The three strands of Tillich's theory of religious symbols. *The Journal of Religion, 46*(1), 104–130.
28. Gavezzotti, A. (2016). Comparing the strength of covalent bonds, intermolecular hydrogen bonds and other intermolecular interactions for organic molecules: X-ray diffraction data and quantum chemical calculations. *New Journal of Chemistry, 40*(8), 6848–6853.
29. Gillespie, R. J., & Robinson, E. A. (2007). Gilbert N. Lewis and the chemical bond:

The electron pair and bond. *Canadian Journal of Chemistry, 85*(12), 1475–1481.

30. Goddard, W. A., & Harding, L. B. (1978). The description of chemical bonding from ab initio calculations. *Annual Review of Physical Chemistry, 29*(1), 363–396.
31. Goncharuk, V. V., Lapshin, V. B., Burdeinaya, T. N., Pleteneva, T. V., Chernopyatko, A. S., Atamanenko, I. D., & Syroeshkin, A. V. (2011). Physicochemical properties and biological activity of the water depleted of heavy isotopes. *Journal of Water Chemistry and Technology, 33*, 8–13.
32. Gordy, W. (1946). A relation between bond force constants, bond orders, bond lengths, and the electronegativities of the bonded atoms. *The Journal of Chemical Physics, 14*(5), 305–320.
33. Grandbois, M., Beyer, M., Rief, M., Clausen-Schaumann, H., & Gaub, H. E. (1999). How strong is a covalent bond? *Science, 283*(5408), 1727–1730.
34. Grant, I. P., & Quiney, H. M. (1988). Foundations of the relativistic theory of atomic and molecular structure. In *Advances in Atomic and Molecular Physics* (Vol. 23, pp. 37–86). Academic Press.
35. Guo, H., Peng, Q., Chen, X. K., Gu, Q., Dong, S., Evans, E. W., & Li, F. (2019). High stability and luminescence efficiency in donor–acceptor neutral radicals not following the Aufbau principle. *Nature Materials, 18*(9), 977–984.
36. Harrowfield, J. (1996). π Stacking and the co-ordinate bond: Sometimes conflicting factors in molecular recognition, as revealed in the structures of metal picrates. *Journal of the Chemical Society, Dalton Transactions*, (15), 3165–3171.
37. Hautier, G., Fischer, C., Ehrlacher, V., Jain, A., & Ceder, G. (2011). Data mined ionic substitutions for the discovery of new compounds. *Inorganic Chemistry, 50*(2), 656–663.
38. Hinze, J., & Jaffe, H. H. (1962). Electronegativity. I. Orbital electronegativity of neutral atoms. *Journal of the American Chemical Society, 84*(4), 540–546.
39. Horton, T. D. (1939). Sinclair Lewis: The symbol of an era. *The North American Review, 248*(2), 374–393.
40. Hughes, R. I. G. (1990). The Bohr atom, models, and realism. *Philosophical Topics, 18*(2), 71–84.
41. Iozzi, M. F., Helgaker, T., & Uggerud, E. (2009). Assessment of theoretical methods for the determination of the mechanochemical strength of covalent bonds. *Molecular Physics, 107*(23, 24), 2537–2546.
42. Jeffries, C. M., Pietras, Z., & Svergun, D. I. (2020). The basics of small-angle neutron scattering (SANS) for new users of structural biology. In *EPJ Web of Conferences* (Vol. 236, pp. 3001). EDP Sciences.
43. Jonsson, M., Lind, J., Merényi, G., & Eriksen, T. E. (1994). Remote substituent effects on polar and non-polar covalent bonds. *Journal of the Chemical Society, Perkin Transactions 2*, (10), 2149–2154.
44. Karol, P. J., Barber, R. C., Sherrill, B. M., Vardaci, E., & Yamazaki, T. (2016). Discovery of the element with atomic number Z= 118 completing the 7th row of the periodic table (IUPAC Technical Report). *Pure and Applied Chemistry, 88*(1, 2), 155–160.

45. Kleitman, D., & Markowsky, G. (1975). On Dedekind's problem: The number of isotone Boolean functions. II. *Transactions of the American Mathematical Society, 213*, 373–390.

46. Kragh, H. (1979). Niels Bohr's second atomic theory. *Historical Studies in the Physical Sciences, 10*, 123–186.

47. Leone, M., Robotti, N., & Verna, G. (2018). 'Rutherford's experiment' on alpha particles scattering: The experiment that never was. *Physics Education, 53*(3), 35003.

48. Ling, T., Jaroniec, M., & Qiao, S. Z. (2020). Recent progress in engineering the atomic and electronic structure of electrocatalysts via cation exchange reactions. *Advanced Materials, 32*(46), 2001866.

49. Lipscomb, W. N. (1966). Framework rearrangement in boranes and carboranes: Cooperative atomic rearrangements are expected in many polyhedron-like electron-deficient molecules and ions. *Science, 153*(3734), 373–378.

50. Mahmood, A., Irfan, A., Ahmad, F., & Janjua, M. R. S. A. (2021). Quantum chemical analysis and molecular dynamics simulations to study the impact of electron-deficient substituents on electronic behavior of small molecule acceptors. *Computational and Theoretical Chemistry, 1204*, 113387.

51. McKagan, S. B., Perkins, K. K., & Wieman, C. E. (2008). Why we should teach the Bohr model and how to teach it effectively. *Physical Review Special Topics—Physics Education Research, 4*(1), 010103.

52. Meek, T. L., & Garner, L. D. (2005). Electronegativity and the bond triangle. *Journal of Chemical Education, 82*(2), 325.

53. Metz, I. K., Bennett, J. W., & Mason, S. E. (2021). Examining the Aufbau principle and ionization energies: A computational chemistry exercise for the introductory level. *Journal of Chemical Education, 98*(12), 4017–4025.

54. Niaz, M. (1998). From cathode rays to alpha particles to quantum of action: A rational reconstruction of structure of the atom and its implications for chemistry textbooks. *Science Education, 82*(5), 527–552.

55. Nordholm, S., & Bacskay, G. B. (2020). The basics of covalent bonding in terms of energy and dynamics. *Molecules, 25*(11), 2667.

56. Owen, J., & Thornley, J. H. M. (1966). Covalent bonding and magnetic properties of transition metal ions. *Reports on Progress in Physics, 29*(2), 675.

57. Parkin, G. (2023). A convenient approach for the construction of Lewis structures by focusing on the valence requirements of outer atoms. *Journal of Chemical Education, 100*(12), 4644–4652.

58. Pauling, L. (1932). The nature of the chemical bond. IV. The energy of single bonds and the relative electronegativity of atoms. *Journal of the American Chemical Society, 54*(9), 3570–3582.

59. Quayle, P. C., Blanton, E. W., Punya, A., Junno, G. T., He, K., Han, L., & Kash, K. (2015). Charge-neutral disorder and polytypes in heterovalent wurtzite-based ternary semiconductors: The importance of the octet rule. *Physical Review B, 91*(20), 205207.

60. Radom, L., Hariharan, P. C., Pople, J. A., & Schleyer, P. V. R. (1973). Molecular

orbital theory of the electronic structure of organic compounds. XIX. Geometries and energies of C_3H_5 cations. Energy relations among allyl, vinyl, and cyclopropyl cations. *Journal of the American Chemical Society, 95*(20), 6531–6544.

61. Reed, A. E., & Schleyer, P. V. R. (1990). Chemical bonding in hypervalent molecules. The dominance of ionic bonding and negative hyperconjugation over d-orbital participation. *Journal of the American Chemical Society, 112*(4), 1434–1445.
62. Ren, J., & Meng, S. (2006). Atomic structure and bonding of water overlayer on Cu(110): The borderline for intact and dissociative adsorption. *Journal of the American Chemical Society, 128*(29), 9282–9283.
63. Schwarz, W. H. (2013). 100th anniversary of Bohr's model of the atom. *Angewandte Chemie International Edition, 52*(47), 63–70.
64. Siegel, D. J., Hector Jr, L. G., & Adams, J. B. (2002). Adhesion, atomic structure, and bonding at the Al(111)/α–Al_2O_3(0001) interface: A first principles study. *Physical Review B, 65*(8), 085415.
65. Smith, D. W. (2007). A new approach to the relationship between bond energy and electronegativity. *Polyhedron, 26*(2), 519–523.
66. Sproul, G. (2001). Electronegativity and bond type: Predicting bond type. *Journal of Chemical Education, 78*(3), 387.
67. Tse, Y. C., Kirkegaard, K., & Wang, J. C. (1980). Covalent bonds between protein and DNA. Formation of phosphotyrosine linkage between certain DNA topoisomerases and DNA. *Journal of Biological Chemistry, 255*(12), 5560–5565.
68. Uce, M. (2015). Constructing models in teaching of chemical bonds: Ionic bond, covalent bond, double and triple bonds, hydrogen bond and molecular geometry. *Educational Research and Reviews, 10*(4), 491.
69. Verma, A. S. (2015). Extant ionic charge theory for bond orbital model based on the tight-binding method: A semi-empirical model applied to wide-bandgap II-VI and III-V semiconductors. *Materials Science in Semiconductor Processing, 29*, 2–15.
70. Wang, Y., Hamers, R. J., & Kaxiras, E. (1995). Atomic structure and bonding of boron-induced reconstructions on Si(001). *Physical Review Letters, 74*(3), 403.
71. Wojciechowski, P. M., Zierkiewicz, W., Michalska, D., & Hobza, P. (2003). Electronic structures, vibrational spectra, and revised assignment of aniline and its radical cation: Theoretical study. *The Journal of Chemical Physics, 118*(24), 10900–10911.
72. Xu, M., Cheng, Y. Q., Sheng, H. W., & Ma, E. (2009). Nature of atomic bonding and atomic structure in the phase-change $Ge_2Sb_2Te_5$ glass. *Physical Review Letters, 103*(19), 195502.
73. Xu, Y. A., Takahashi, K., Goriely, S., Arnould, M., Ohta, M., & Utsunomiya, H. (2013). NACRE II: An update of the NACRE compilation of charged-particle-induced thermonuclear reaction rates for nuclei with mass number $A < 16$. *Nuclear Physics A, 918*, 61–169.
74. Yock, P. (2021). Comparison of Rutherford's atomic model with the Standard Model of particle physics and other models. *Journal of the Royal Society of New Zealand, 51*(3, 4), 538–556.

75. Zeng, Y., Zou, R., Luo, Z., Zhang, H., Yao, X., Ma, X., & Zhao, Y. (2015). Covalent organic frameworks formed with two types of covalent bonds based on orthogonal reactions. *Journal of the American Chemical Society, 137*(3), 1020–1023.

76. Zhan, C. G., Nichols, J. A., & Dixon, D. A. (2003). Ionization potential, electron affinity, electronegativity, hardness, and electron excitation energy: Molecular properties from density functional theory orbital energies. *The Journal of Physical Chemistry A, 107*(20), 4184–4195.

CHAPTER 3

Inorganic Chemistry

LEARNING OBJECTIVES

After studying this chapter, you will be able to:

1. Understand the fundamentals of inorganic chemistry, including its applications in organometallic chemistry, environmental implications, and industrial applications.
2. Explain the nomenclature of inorganic compounds according to IUPAC rules, including naming compounds with different charged states and transition metals.
3. Classify inorganic chemicals based on their physical and chemical properties, types of bonds, reactivity, and valence.
4. Analyze the significance of isotopes and nuclides in inorganic chemistry.
5. Describe the classification and properties of minerals.
6. Discuss the role of the periodic table in organizing elements and predicting their properties.
7. Comprehend the equilibrium constant in relation to inorganic chemical reactions.

KEY TERMS FROM THIS CHAPTER

Environmental implications
Ionic compounds
Minerals
Polyatomic ions
Equilibrium constant
Ionic liquids
Organometallic chemistry

INTRODUCTION

If organic chemistry is defined as the chemistry of hydrocarbon compounds and their derivatives, inorganic chemistry can be described very generally as the chemistry of noncarbon compounds or as the chemistry of everything else. This includes all the remaining elements in the periodic table and some compounds of carbon (such as carbon monoxide (CO) and carbon dioxide (CO_2)), which plays a major role in many inorganic compounds. Thus, inorganic chemistry is the subcategory of chemistry concerned with the properties and reactions of inorganic compounds, which includes all chemical compounds without the chains or rings of carbon atoms that fall into the subcategory of organic compounds (Mjos and Orvig, 2014).

Inorganic chemistry is a distinct field of chemistry that focuses on comprehending the characteristics, arrangements, and interactions of substances that do not contain carbon chains or rings. These compounds, commonly known as inorganic substances, encompass a wide range of materials, including salts, oxides, and coordination complexes. In contrast to organic compounds, which rely on carbon-hydrogen structures, inorganic compounds frequently incorporate elements from various parts of the periodic table, showcasing a wide range of chemical properties and practical uses (Benkhay et al., 2020) (Figure 3.1).

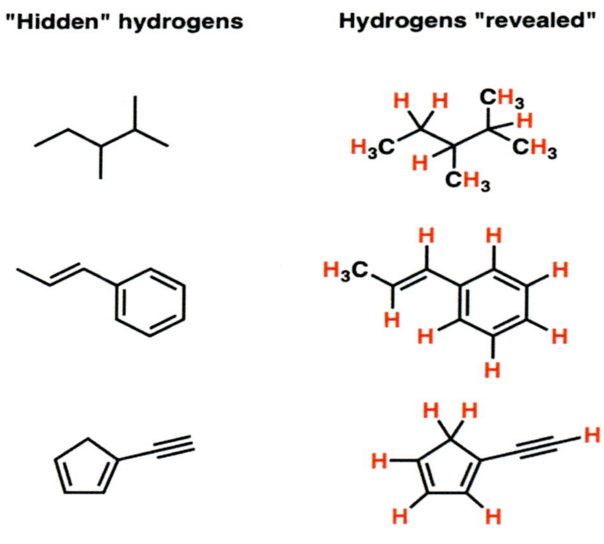

Figure 3.1. *Figure of carbon and hydrogen structure.*

Source: Master organic chemistry, Creative Commons License.

Inorganic chemistry plays a vital role in numerous industrial processes and technological breakthroughs. It encompasses the examination of metals and their alloys, which are vital for the fields of construction, manufacturing, and electronics. Inorganic chemists investigate the functions of minerals and materials in several applications, including catalysis, energy storage, and environmental remediation. Moreover, a comprehensive detail of the concepts of inorganic chemistry is essential for grasping the manner in which elements and their compounds behave in natural systems, including everything from the Earth's crust to biological creatures (Josephsen and Kristensen, 2006).

Inorganic chemistry provides a broad perspective of the chemical world that goes beyond organic substances. It offers valuable insights into the wide range of chemical interactions and the various applications of inorganic materials in research and business (Kaltsoyannis and McGrady, 2004).

3.1. FUNDAMENTALS OF INORGANIC CHEMISTRY

An effective method to differentiate between inorganic and organic compounds is based on the origin of these substances. Inorganic compounds are formed either through natural processes that are not associated with any living organisms or through deliberate human experimentation in a laboratory setting. On the other hand, organic compounds are produced as a result of the biological activity of living beings. Nevertheless, it is prudent to use caution while employing this definition, as organic substances can be synthetically produced in a laboratory setting. Another description refers to the salinity-producing characteristic of inorganic substances, which is not present in organic compounds. It is important to note that even this definition is not completely true because organic acids (RCO_2H) are also capable of producing salts.

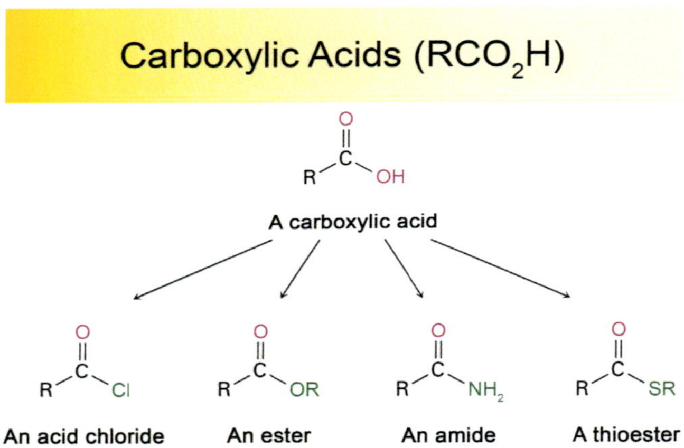

Figure 3.2. *Schematic of carboxylic acid.*

Source: Slideplayer, Creative Commons License.

The absence of carbon-hydrogen bonds, which are a distinguishing feature of organic compounds, is another argument that can be made against inorganic substances (Figure 3.2). On the other hand, this argument is not totally correct because perfluorocarbons, which are compounds in which all hydrogen atoms are replaced by fluorine atoms, do not contain bonds between carbons and hydrogen, yet they are nevertheless considered to be organic compounds. Another discrepancy that is commonly brought out is the fact that organic molecules do not include any metal atoms, whereas inorganic entities do carry metal atoms. To reiterate, this assertion is not true because organometallic compounds are made up of atoms of metal. Therefore, it is essential to exercise caution when trusting any definition that purports to differentiate between inorganic chemicals and organic molecules. This is because it is important to identify the differences between the two (Murphy et al., 1998) (Figure 3.3).

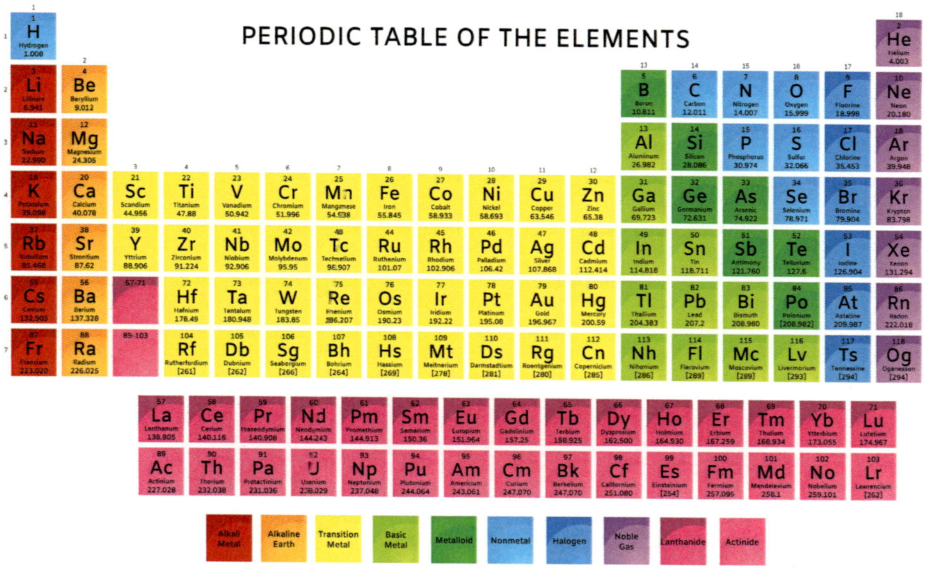

Figure 3.3. *Image of periodic table.*

Source: https://sciencenotes.org/wp-content/uploads/2020/04/PeriodicTableAtomicMassColor.png.

3.1.1. Organometallic Chemistry

Organometallic chemistry is a vast and fast-expanding discipline that focuses on molecules with direct metal-carbon bonds. It also encompasses the study of catalysis in many chemical reactions. Organometallic compounds consist of a minimum of one link connecting a metallic atom with a carbon atom. These compounds are referred to as coordination compounds (Ernst, 1999). Additionally, environmental chemistry encompasses the examination of both inorganic and organic substances. Therefore, the exploration of these distinct branches of chemistry is crucial for a comprehensive understanding of the subject. The field of inorganic chemistry is vast, offering countless opportunities for research and exploration (Bertrand and Casini, 2014).

In the most inclusive sense, inorganic materials and compounds can be described by what they lack:

i. They are not of organic origin, and,

ii. Any substances that extend beyond living, hydrocarbons, and other carbon-based chemicals can be classified as inorganic (Feher and Budzichowski, 1995).

Inorganic chemicals are solid compounds that lack carbon in their structure as molecules and are mostly composed of the most prevalent elements on earth, such as silicon, oxygen, aluminum,

Remember

Remember the fundamental principles of inorganic chemistry lay the groundwork for understanding various chemical interactions and applications in both natural and industrial contexts.

calcium, iron, salt, potassium, and magnesium. The only exceptions are carbon monoxide and carbon dioxide together with derivatives of the mineral carbonate and bicarbonate (Hoffmann, 1981).

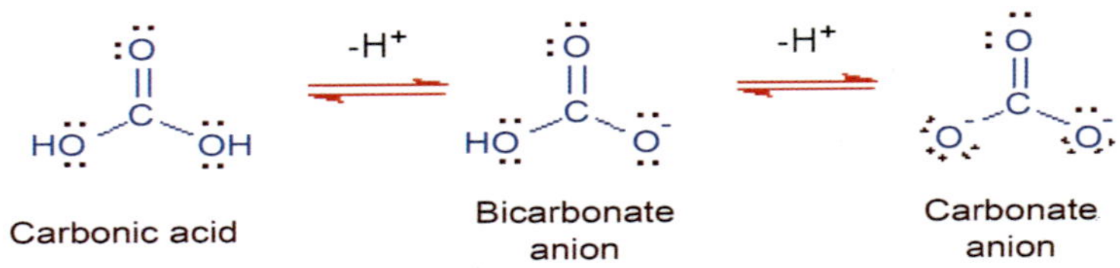

Figure 3.4. *Image of molecular structure of carbonate and bicarbonate.*

Source: Adi chemistry, Creative Commons License.

3.1.2. Environmental Implications

Regarding environmental concerns about spills or elimination of inorganic chemicals, there is already a significant amount of knowledge regarding how the molecular structure of chemicals affects their toxicity to humans (Bondu et al., 2021). However, there is much less understanding about how the molecular structure of a chemical influences its ability to endure in the environment. In ecosystems, the longevity of chemicals is a crucial factor in determining possible harm.

This is because there are undoubtedly certain species that are highly sensitive to any chemical, especially those that persist over time (Su, 2017). While certain chemicals may not pose a threat to a small number of creatures, they will ultimately be transported through biogeochemical cycles and finally reach a vulnerable species within an ecosystem. Therefore, compounds that are highly hazardous and easily broken down by natural processes may provide a smaller environmental concern compared to a relatively harmless chemical that persists in the ecosystem and has the potential to harm important plant or animal species (Slaveykova et al., 2020).

3.1.3. Chemical Effects in the Environment

Hence, the examination of chemical impacts on the environment can be divided into two areas of study:

1. An investigation into the concentrations of a substance that accumulate in the air, water, soils, and living organisms, and
2. An examination of the consequences that occur in living organisms when the threshold level of action has been reached, specifically the adverse effects that are observed (Arnot et al., 2006).

To accurately forecast changes in the levels of a chemical, further data is required regarding the injection rates, flow dynamics, distribution among air, water, soils, and biota, as well as the extent of degradation, which leads to the development of an ecological balance sheet for an environment. The behavior of these dynamic processes is determined by both the chemical and physical characteristics of the molecular impurity (Bai and Kohlstedt, 1993).

In the future, fluid dynamics and meteorology could potentially offer the necessary conceptual and technical resources to develop predictive models for these systems. The majority of current understanding regarding the impacts of substances comes from severe toxicology and medical research conducted on a range of organisms, including humans. However, as environmental effects are typically linked to prolonged exposure, there is a growing emphasis on studying the long-term effects of continuous, low-level exposure to chemicals (Sharma and Singhvi, 2017). The challenge of identifying these effects in a biological system is made worse by the fact that plenty of them are general and can be easily confused with similar effects caused by natural events like famines, droughts, or various meteorological or disastrous phenomena. When a real impact is acknowledged, it is necessary to identify and establish a connection between the effect and a potential cause (Kumar et al., 2019). Experimental research, including laboratory and field work, is necessary to establish a clear link between certain inorganic chemicals and any negative impacts on the ecosystem. This link should be established beyond any reasonable question. Only through diligent gathering and incorporation of technical knowledge in the field of inorganic chemistry, specifically about the characteristics and actions of inorganic compounds, can this be accomplished (Connon et al., 2012).

Nevertheless, it is not anticipated that an engineer will amass an equivalent level of chemical expertise as an experienced inorganic chemist. Similarly, the chemist may feel uncomfortable or hesitant about having to acquire proficiency in one or more engineering fields. However, the process involves acquiring enough knowledge to comprehend the conduct of inorganic substances in the environment and then being able to make a credible prediction about their behavior based on their attributes. The lack of acknowledgment of the interconnected responsibilities of chemists and engineers would impede and harm the progress of a cohesive sustainable development policy that may be applied to ensure the long-term viability of any ecosystem or large environmental region (Akingbemi and Hardy, 2001).

Did you know?

Did you know that inorganic chemicals can have significant impacts on the environment, influencing ecosystems and human health through processes like acid rain formation and contamination of water sources?

3.1.4. Industrial Applications

Inorganic chemicals find utility in various domains of the chemical business, encompassing catalysts, colorants, surfactants, coverings, medical care, fuel, and farming. The outcomes of inorganic chemical processes serve as:

1. Fundamental substances for industrial procedures, encompassing bases, acids, mineral salts, oxidizers, gases, and halogens,
2. Chemical enhancements, comprising colorants, alkali metals, and
3. End products, such as chemical fertilizers, building glass, and building materials.

From an industrial standpoint, there are two primary categories of inorganic chemicals:

1. Alkali chemicals, which consist of soda ash $NaCO_3$, caustic soda NaOH, and Cl_2, and
2. Basic inorganic substances which include AlF_3, $KClO_3$, CaC_2, and TiO_2.

Moreover, the chlor-alkali (Figure 3.5) business plays a significant role in the global chemical economy. An important industrial reaction is the electrolysis of brine, which contains NaCl, to produce NaOH, Cl_2, and H_2 (Vallarino and Wnek, 2002).

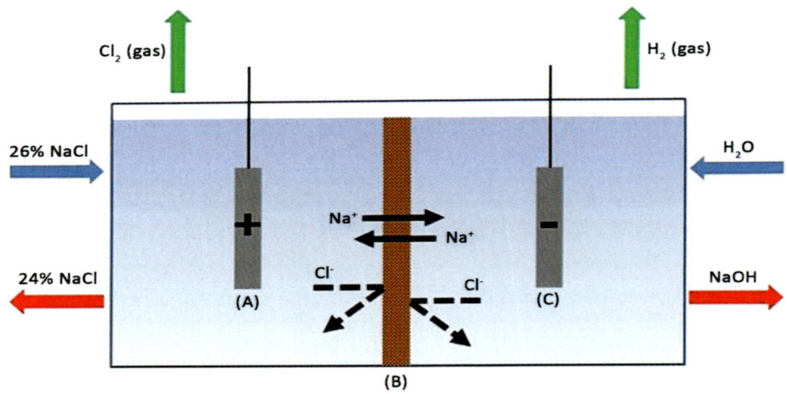

Figure 3.5. *Illustration of Chlor-Alkali process.*

Source: Lenntech, Creative Commons License).

3.2. NOMENCLATURE OF INORGANIC COMPOUNDS

Establishing a universally accepted chemical nomenclature is crucial for effective communication within the field of chemical sciences as well as regulatory purposes, including commercial activities and environmental health and safety. Names and formulas, also known as formulae in Latin, fulfill just half of their purpose when they are generated and utilized for describing or recognizing compounds, such as in publications. In order to completely fulfill their purpose, the reader must possess the ability to accurately comprehend a given name or formula, enabling them to, for instance, generate a structure diagram (Jorissen et al., 1941).

3.2.1. International Union of Pure and Applied Chemistry (IUPAC)

The International Union of Pure and Applied Chemistry (IUPAC) recommends a system of naming and nomenclature for inorganic chemicals. In this system, the names of ionic compounds are based on their content rather than their structure. This enables the identification of a chemical compound to be universally acknowledged by its name, regardless of the language spoken by the chemist. For a binary ionic substance with clear charges and stoichiometry, its general term is expressed as a two-word phrase (McCusker et al., 2001). The cation, which is the unmodified name of the element for monatomic cations, is listed initially, preceded by a description of the anion. As an illustration, the chemical compound $MgCl_2$ is officially designated as magnesium chloride, while Na_2SO_4 is officially designated as sodium sulfate. Furthermore, the sulfate ion (SO_4^{2-}) serves as an exemplification of a polyatomic ion, which refers to an ion composed of several atoms (Alemán et al., 2007).

3.2.2. Multiplicative Prefixes

When dealing with multiple cations and/or anions, it becomes necessary to use multiplicative prefixes to indicate the relative compositions. The cations are listed first, followed by the anions, in alphabetical order. Both the empirical formula and the systematic name reflect cations in alphabetical order. If one of the ions already includes a multiplicative prefix in its name, alternative prefixes like bis-, tris-, and tetrakis- are used for the others. For example, the chemical formula $Ba(BrF_4)_2$ is fully expressed as barium bis(tetrafluoride bromate) (McCleverty and Connelly, 2001).

3.2.3. Compounds with Different Charged States

For the sake of preserving the molecule's overall neutrality, compounds that contain elements that are capable of residing in a variety of oxidation states will display a stoichiometry that is defined by the oxidation state. This may be communicated through the name by referring to either the degree of oxidation that the elements are in or the charge that is imposed on the ions. Due to the possibility of uncertainty in assigning oxidation states, IUPAC prefers to indicate the ionic charge values (Lecomte et al., 2005). The cations are denoted by Arabic numerals that are followed by the sign (2, 1, 1+, and 2+). These Arabic integers are put in parentheses right after the cation's name, and there is no space between them.

As an example, the molecule $FeSO_4$ is known as ferrous sulfate. This is because the iron ion, which is denoted by the symbol Fe^{2+}, possesses a charge of 2+, which helps to counteract the charge of 2– on the sulfate ion. Ferric sulfate, on the other hand, is a compound that is known as $[Fe_2(SO_4)_3]$. This is due to the fact that each of the two ions of iron in the formula has a 3+ charge, which helps to balance the 2– charge that is present on all three sulfate ions. In spite of this, the expression of the oxidation number using Roman numbers is still considered to be the standard approach (as demonstrated in this article). In accordance with the preceding illustration, the compounds would be designated as iron(II) sulfate, which would then be followed by iron(III) sulfate (Mulliken, 1952). Whenever it comes to simple ions, the charge of the ion and the oxidation number are identical to one another. Polyatomic ions, on the other hand, frequently have values that are distinct from one another.

A naming system that is more ancient, which is usually referred to as the classical system, is still in use for metal cations and is still in existence. In order to give distinct names for the lower and higher oxidation states, respectively, this approach requires the addition of the suffixes -ous and -ic to the Latin root of the word. Once more, using the compounds stated above as an illustration, this way of naming uses the terms "ferrous" and "ferric" (which are derived from the Latin word for iron, "ferrum") to denote iron(II) and iron(III) respectively. As a consequence of this, the examples that were discussed earlier would be referred to as ferrous sulfate and ferric sulfate, which is a nomenclature that is still frequently utilized. Chemists continue to have a tendency to use the Roman and Arabic forms of naming, despite the fact that the methodology that is supported by IUPAC has been adopted on a global scale (Krivovichev et al., 2007).

3.2.4. Ionic Compounds

Chemical compounds composed of a metal and a nonmetal are commonly referred to as ionic compounds, and their names typically end in -ide. Cations carry positive charges, while anions carry negative charges (*see* Tables 3.1 and 3.2), and their names are derived from their parent metal (Table 3.3). Transition metals can form multiple ions. To specify which ion is being referred to, a Roman numeral is assigned after the metal's name. This numeral represents both the electrical charge and the oxidation state of the metal. For further details, please refer to Table 3.4 (Nyquist and Kagel, 2012).

Table 3.1. *Tabular Representation of Cation and Anion*

(Source: Speight 2017, Creative Commons License).

+1 Charge	+2 Charge	21 Charge	22 Charge 23 Charge		24 Charge
Hydrogen: H+ Sodium: Na^+ Lithium: Li^+ Rubidium: Rb^+ Potassium: K^+ Cesium: Cs^+	Calcium: Ca^{2+} Beryllium: Be^{2+} Strontium: Sr^{2+} Magnesium: Mg^{2+} Barium: Ba^{2+}	Chloride: Cl			
		Fluoride: F	Sulfide: S_2	Phosphide: P_3	Carbide: C_4
		Hydride: H	Oxide:O_2	Nitride: N_3	
		Bromide: Br			
		Iodide: I			

Table 3.2. *Tabular representation of Transition Metal and Metal Cations*

(Source: Speight 2017, Creative Commons License).

+1 Charge	+2 Charge	+3 Charge	+4 Charge
Silver: Ag^+	Copper(II): Cu^{2+}	Cobalt(III): Co^{3+}	Lead(IV): Pb^{4+}
Copper(I): Cu^+	Cobalt(II): Co^{2+} Tin(II): Sn^{2+} Nickel: Ni^{2+} Lead(II): Pb^{2+} Zinc: Zn^{2+}	Iron(III): Fe^{3+}	Tin(IV): Sn^{4+}
	Iron(II): Fe^{2+}	Aluminum: Al^{3+}	

Table 3.3. *Tabular representation of Nomenclature of Metals Ions*

(Source: Speight 2017, Creative Commons License).

Metal Io^{n}a	Latin Name
Lead (IV): Pb^{4+}	Plumbic

Lead (II): Pb^{2+}	Plumbous
Iron (III): Fe^{3+}	Ferric
Iron (II): Fe^{2+}	Ferrous
Tin (IV): Sn^{4+}	Stannic
Tin (II): Sn^{2+}	Stannous
Mercury (II): Hg^{2+}	Mercuric
Mercury (I): Hg_2^{2+}	Mercurous
Copper (II): Cu^{2+}	Cupric
Copper (I): Cu^{+}	Cuprous

An ionic substance must have a net charge of zero, indicating that it is electrically neutral. As an illustration, a single Na^{+} is combined with a single Cl^{-}, while a lone Ca^{2+} is matched with two ions of bromide (Br) (Kaya and Kaya, 2015).

3.2.5. Transition Metals

Certain charges exhibited by transition metals (Figure 3.6) are assigned special Latin names. Similar to other laws of nomenclature, the transition metal ion with the lower charge is given a Latin name ending in -ous, while the one with the greater charge is given a Latin name ending in -ic. Nevertheless, there are a few exemptions to the assignment of Roman numerals, such as zinc, aluminum, and silver. Despite being classified as transition metals, these elements do not require Roman numbers to indicate their oxidation state in their names, as they only exist as a single ion. Instead of employing Roman numerals, the various ions can alternatively be expressed using ordinary words. The metal undergoes a transformation, resulting in the addition of either -ous or -ic at the end (Dobbs and Hehre, 1987).

HF, also known as hydrogen fluoride, is specifically referred to as an acid due to its ability to dissociate into ions of hydrogen (H+) and anions when dissolved in water. An expedient method for identifying acids is to check for the presence of an H (hydrogen) preceding the chemical formula of the substance. In order to designate acids, the letter hydro- is added before the nonmetal which is altered to have the suffix -ic. Acids exist in an aqueous state (aq) due to their presence in water. Several frequently encountered binary acids include (Boehme et al., 2000):

$$HF(g)(hydrogen\ fluoride) \rightarrow HF(aq)(hydrofluoric\ acid)$$

$$HBr(g)(hydrogen\ bromide) \rightarrow HBr(aq)(hydrobromic\ acid)$$

$$HCl(g)(hydrogen\ chloride) \rightarrow HCl(aq)(hydrochloric\ acid)$$

$$H_2S(g)(hydrogen\ sulfide) \rightarrow H_2S(aq)(hydrosulfluric\ acid)$$

Figure 3.6. *Illustration of transition metals.*

(Source: Jack Westin, Creative Commons License).

3.2.6. Polyatomic Ions

Polyatomic ions, which consist of two or more atoms, are connected through covalent bonds (as shown in Table 3.4) (Erlander, 1968).

Tip

Polyatomic ions are clusters of atoms that carry a net electrical charge and play essential roles in chemical reactions and compound stability. Understanding their structures and charges aids in predicting their behavior in chemical reactions.

Table 3.4. *Tabular representation of Common Polyatomic Ions*

(Source: Speight, J.G., Creative Commons License).

Cation Anion (ion)	Formula
Ammonium	NH4+
Arsenate	AsO_4^{-3}
Acetate	$C_2H_3O_2$
Hypochlorite	ClO^-
Hydronium	H_3O^+
Carbonate	CO_3^{2-}
Chromate	CrO_4^{2-}
Chlorite	ClO_2^-
Hydroxide	OH^-
Perchlorate	ClO_4^-
Chlorate	ClO_3^-
Dichromate	$Cr_2O_2^{-7}$
Cyanide	CN^-
Oxalate	$C_2O_4^{2-}$
Sulfate	SO_4^{2-}
Nitrate	NO_3^-
Phosphate	$[PO_4]^{3-}$
Permanganate	MnO_4^-
Nitrite	NO^{-2}
Sulfite	SO_3^{2-}
Thiosulfate	$S_2O_2^{-3}$
Thiocyanate	SCN

A hydroxide ion (OH) is an instance of a polyatomic ion, composed of one atom of oxygen and one hydrogen atom, with a charge of 1. In a comparable manner an ammonium ion (NH_4^+) consists of a single nitrogen atom and 4 hydrogen atoms, giving a charge of +1.

A polyatomic ion can be conceptualized as the conjugated acid/base to a neutral molecule in many cases. As an illustration, HSO_4 is the conjugate base of sulfuric acid (H_2SO_4). While the hydrogen ion (H^+) may possess a positive charge, it does not form an ionic bond with another element. This phenomenon arises as once the atoms have already created an ionic bond, they have already transformed into a compound and hence do not require any further gain or loss of electrons (Gibson, 1997).

Another category of inorganic substances that has recently appeared is ionic liquids (Table 3.5). These are:

Table 3.5. *Tabular Representation of General Properties of Ionic Liquids*

Property	Comment
A salt	Anion is small
A salt	Cation is large
Freezing point	Below 100°C
Melting point	Below 100°C
Thermal stability	High
Liquid range	>200°C
Dielectric constant	<30
Viscosity	<100 cP
Solvent properties	Good
Polarity	Moderate
Vapor pressure	Low to negligible
Catalytic properties	Good

(*Source:* Emsley, Creative Commons License).

3.2.7. Ionic Liquids

Most liquids are mostly electrically neutral, whereas ionic liquids consist mainly of ions and temporary ion pairs. These compounds are commonly referred to by several names such as electrolytes, ion melts, ionic liquids, amalgamated salts, fluid salts, or ionic glass. Ionic liquids are highly efficient solvents and electrically conductive fluids, sometimes known as electrolytes. Any salt that undergoes melting without undergoing decomposition or vaporization typically produces an ionic liquid. On the other hand, whenever an ionic liquid cools down, it typically transforms into an ionic solid, which can have either a crystalline or glass-like structure (Gao et al., 2017).

3.2.8. Isotopes

An isotope represents a unique variant of a chemical element distinguished by varying numbers of neutrons in its atomic structure. While each isotope of a specific element maintains an identical count of protons within its atoms, multiple isotopes of the same element are positioned in the corresponding location on the periodic table. The atomic number of an atom, denoting the number of protons in its nucleus, equals the number of electrons in a neutral, non-ionized state. This atomic number serves to uniquely identify an element but not its specific isotopes. Isotopes of an element can exhibit a range of neutron counts (Bartholomay et al., 1996).

3.2.9. Nuclide

The term "nuclide" specifically refers to the nucleus of an atom, distinct from the atom as a whole. However, "isotope," though older, is more widely recognized and sometimes used in contexts where "nuclide" would be more appropriate. A nuclide is characterized by identical nuclei; for example, carbon-13 consists of nuclei each containing 6 protons and 7 neutrons. Nuclide concepts primarily emphasize nuclear properties rather than chemical properties, whereas isotopes categorize atoms of each element based on chemical characteristics rather than nuclear characteristics (Fernelius et al., 1979).

3.3. CLASSIFICATION OF INORGANIC CHEMICALS

In terms of their molecular composition, inorganic chemicals can be classified into two categories:

i. Simple inorganic compounds and,

ii. Complicated inorganic chemicals.

Examples of simple inorganic substances are molecules comprised solely of atoms of a single element. These molecules are unable to be broken down into other chemicals by chemical processes. In contrast, complex inorganic substances are formed of diverse atoms, representing many chemical elements. During chemical reactions, these molecules undergo decomposition, resulting in the production of multiple additional chemicals (Kirchhübel and Fantke, 2019). Therefore,

Simple	**Metals** **Nonmetals**
Complex	Acids Bases Oxides Salts

The boundary between metals and nonmetals can often be blurred, as many substances exhibit characteristics that straddle both categories. Inorganic chemicals are broadly categorized into two main groups based on their composition: ionic and molecular compounds. The systematic nomenclature of these compounds ensures clear identification and differentiation as distinct substances (Verdonck et al., 2017).

In a broader context, inorganic substances exclude biological molecules and predominantly consist of non-carbon-based chemicals, encompassing bioinorganic substances, hydrocarbon derivatives, and other carbon-based chemicals that share similarities. The simplicity of inorganic compounds is often attributed to their absence of carbon bonds. These compounds are derived from chemical elements organized according to the periodic table (Figure 2.1, Tables 2.7 and 2.8) and are frequently sourced from minerals (Tables 2.9 and 2.10). They are primarily composed of abundant earthly elements such as silicon, iron, calcium, sodium, oxygen, potassium, and magnesium (Vassilev et al., 2009).

Inorganic synthesis involves the production of various fundamental chemical molecules that lack carbon. For example, inorganic pigments consist of metallic oxides, sulfur dioxide, or other salts occurring naturally or synthesized. These pigments undergo calcination during manufacturing at temperatures ranging from 1200 to 2100°F (650 to 1150°C). Known for their exceptional thermal stability, resistance to light degradation, weathering resilience, and resistance to migration, inorganic pigments find extensive applications in electroplating, dyeing, precision casting, alcohol distillation, and papermaking.

Examples of inorganic compounds used as preservatives include sulfur dioxide and sulfite derivatives. Beyond their preservative roles, inorganic compounds serve as essential components in feed extracts, pesticides, wood preservers, and disinfectants. The distinct characteristics and diverse applications of inorganic chemicals underscore their critical role in various industrial processes and everyday applications (Roschkowski and Simmler, 1980).

3.3.1. Physical and Chemical Properties

Elements can be categorized into three main groups based on their shared chemical and physical properties: metals, metalloids, and nonmetals. Metals are often lustrous, excellent conductors of electricity, and solid elements that may create alloys when combined with other metals, as well as ionic compounds resembling salts when combined with (Patnaik, 2003) (Tables 3.6 and 3.7).

Table 3.6. *Tabular Form of the Periodic Table Showing the Ionic Components*

Group	Description	Examples
Alkali metals	Group 1	Sodium (Na) Lithium (Li) Francium (Fr) Rubidium (Rb) Potassium (K) Cesium (Cs)
Alkali earths	Group 2	Beryllium (Be) Magnesium (Mg) Calcium (Ca) Radium (Ra) Barium (Ba) Strontium (Sr)

Metalloids	Semimetals	Boron (B) Germanium (Ge) Silicon (Si) Arsenic (As) Tellurium (Te) Antimony (Sb) Polonium (Po)
Nonmetals	Poor conductors	Hydrogen (H) Nitrogen (N) Sulfur (S) Carbon (C) Oxygen (O) Phosphorus (P) Selenium (Se)
Transition metals	Groups 3–10	Silver (Ag) Platinum (Pt) Gold (Au) Titanium (Ti) Iron (Fe)
Other metals	Groups 13, 14, and 15	Aluminum (Al) Indium (In) Gallium (Ga) Tin (Sn) Lead (Pb) Thallium (Tl) Bismuth (Bi)
Rare earths	A collection of 17 elements	Scandium Lanthanides Yttrium

(Source: Cotton, F.A, Creative Commons License).

Table 3.7. *Tabular Form of the Periodic Table Showing Anionic Components*

Group	Description	Examples
Halogens	Group 17	Fluorine (F) Bromine (Br) Chlorine (Cl) Astatine (At) Iodine (I)
Nonmetals	Poor conductors having low-density	Hydrogen (H) Carbon (C) Phosphorus (P) Oxygen (O) Nitrogen (N) Sulfur (S) Selenium (Se)
Metalloids	Semimetals	Boron (B) Germanium (Ge) Silicon (Si) Arsenic (As) Tellurium (Te) Antimony (Sb) Polonium (Po)
Oxoanions	Polyatomic anions	Borates ($BO_{3,4}$) Carbonates (CO_3) Bromates (BrO_3) Chlorates (ClO_3) Nitrates (NO_3) Cyanates (NCO) Phosphates (PO_4) Sulfates (SO_4) Silicates (SiO_4)
Other		Hydroxides (OH) Cyanides (CN)

(*Source:* Cotton, F.A, Creative Commons License).

Insulating gasses are either colored or colorless, and the majority of nonmetals also exhibit this quality. When nonmetals join with other nonmetals, they often exhibit covalent bonding, which is characterized by the collaboration of electrons. Metalloids have characteristics that are either transitional or a combination of both of these types of characteristics (Day, 1935).

Subcategories that illustrate a shift from metallic features to nonmetallic traits can be used to classify metals and nonmetals after they have been separated into several categories. In accordance with their level of reactivity, the metals can be divided into a variety of distinct groups. These groups include the highly reactive alkali metals, the less active alkaline earth metallic substances, the lanthanide series (elements ranging from atomic number 59 to 70), the actinide elements (elements ranging from atomic number 89 to 102), and ultimately the post-transition metals, which are characterized by their poor physical and chemical properties. It is possible to further categorize the nonmetals into polyatomic nonmetallic substances. These substances, which are close to the metalloids, exhibit certain metallic qualities due to their proximity to the metalloids. The diatomic nonmetals occupy the next position in the series, which is then followed by the nonmetallic monatomic noble gases, which are characterized by a high degree of inertness. There are a variety of alternative categorization schemes that can be utilized, such as classifying the elements according to their crystalline structure or the mineralogical occurrence of individual elements (Audrieth and Daniels, 1951).

The elements are not the only major categories of inorganic compounds; many other significant categories exist as well. Among these are bioinorganic chemicals, which include both naturally occurring and artificially produced molecules containing metallic elements that are linked to proteins and other physiological chemistries. Cluster compounds, larger than typical molecules, consist of groups of atoms bonded together. Additionally, coordination compounds hold great importance; these involve a central ion, often a transition metal ion, surrounded by an array of anions or charged molecules. Organometallic compounds are another category, characterized by carbon atoms directly bound to a metal ion. Lastly, there are numerous solid-state compounds, which are solid under typical temperatures and pressures. These compounds exhibit several distinctive properties, particularly in the domain of semiconductors (Ji et al., 2010).

Both liquids and gasses, as well as solids, are all possible forms that inorganic compounds can take. The classification of these substances, on the other hand, is not largely determined by their physical state (whether they are gaseous, liquid, or solid), but rather by the information that is gained from their chemical and physical properties (as demonstrated in Table 2.11) (Judeinstein and Sanchez, 1996). Liquids and gases, when subjected to typical circumstances of pressure and temperature, adapt to the shape of the container they are contained in. Solids, on the other hand, have a definite shape and volume and are held together by intense interactions between molecules and between atoms. A great number of solid inorganic substances are characterized by the presence of forces that are sufficiently powerful to maintain the atoms in a distinct and well-organized crystal state. Some solids, on the other hand, do not possess a crystal structure and instead continue to exist in a disordered or amorphous state (Clarke et al., 2008).

The inorganic gases have a lower number of intermolecular connections, which leads to rapid diffusion and the capacity to conform to the shape of their container. On the other hand, the volume of the gas is affected by temperature and pressure, both of which are frequently determined by the dimensions of the container (Sanchez et al., 2003). In contrast, the molecules that make up inorganic fluids are quite close to one another (in comparison to the molecules that make up gases), and the intermolecular forces that exist between molecules tend to keep the volume of the fluid stable at a very particular temperature. The molecular density of inorganic solvents is higher than that of gases, which results in rather strong intermolecular interactions. Inorganic solvents are substances that have a lower molecular density (Kang et al., 2001).

One way to classify inorganic compounds is according to the presence of functional groups inside their molecules. For example, a nitrate group (NO_3) or a sulfate group (SO_4) are examples of functional groups. In most cases, the functional group of an inorganic molecule is the factor that best determines how the substance behaves in its environment (Ravinder et al., 2021). A functional group is a particular set of atoms that are contained within a molecule and that demonstrate consistent reactivity across a variety of molecules. It is anticipated that steric considerations will not hamper the reactivity of the functional group. Despite the fact that functional groups in inorganic chemistry are not frequently explored to the same extent as they are in organic chemistry, they play an essential part in the classification of the structure of inorganic compounds and the prediction of their physical and chemical properties (Kosmulski, 2001).

A collection of atoms or an element itself that is capable of forming a product through a reaction with a reactive species is referred to as a functional group. By virtue of its presence, a functional group possesses the capacity to exert an impact on the properties of the molecule in which it was initially present (Mackenzie and Bescher, 2000). The individual atom or group of atoms that constitute a functional entity within an inorganic molecule is referred to as a functional group. This can be a positively charged ion (cation) or a negatively charged ion (anion), depending on the nature of the ion. The aggregate contribution of all of the atoms that are contained within this functional entity to the overall qualities of the molecule is what distinguishes it from other entities (e.g., as was seen in Table 2.12) (Fadeel and Garcia-Bennett, 2010).

When it comes to the field of chemistry, inorganic compounds can be classified according to a number of different criteria. One of the most popular approaches that is utilized is based on the specific components that they contain, which is tied to the presence of a functional group, but is not always referred to as such. For example, oxides are made up of one or more atoms of oxygen, hydrides are made up of one or more hydrogen atoms, and halogens are made up of one or more hydrogen atoms (Selvi et al., 2005) (Table 3.8).

Table 3.8. *Tabular Representation of Some Common Inorganic Elements and Mineral*

Material/Chemical	Occurrence	Applications
Barium	Barite contains a high concentration of the dense metallic element barium.	Barium sulfate, derived from barite, is used in oil well drilling, the pulp and paper industry, and the rubber industry. It serves as a filler or extender in cloth, dye, and plastic products. In medical radiography, it is used as a contrast agent. Barium sulfate is also employed in the production of high-quality white pigments.
Silicon	Silicon is the second most abundant element in the Earth's crust, primarily found in sand, quartz, and silicate minerals.	Used extensively in electronics as a semiconductor in integrated circuits and solar cells. Silicon is also used to make glass, ceramics, and silicones for lubricants and medical implants.
Aluminum	Aluminum is commonly found in the mineral bauxite, which is the primary source of aluminum.	Used in the production of aircraft, automobiles, and packaging materials like cans and foil. Aluminum is also used in construction materials and electrical applications.
Iron	Iron is predominantly found in minerals like hematite and magnetite.	Used to manufacture steel, which is essential in construction, automotive, and machinery industries. Iron is also used in the production of magnets, tools, and appliances.
Copper	Copper is primarily extracted from minerals such as chalcopyrite and malachite.	Used in electrical wiring, plumbing, and the production of electronics and electrical equipment. Copper is also used in coins, jewelry, and musical instruments.
Zinc	Zinc is commonly found in the mineral sphalerite.	Used to galvanize steel to prevent rusting. Zinc is also used in the production of batteries, alloys, and as a dietary supplement.
Lead	Lead is primarily found in the mineral galena.	Used in the production of batteries, ammunition, and radiation shielding. Lead is also used in the manufacture of glass and ceramics.

Sodium Chloride	Commonly known as salt, it is found in large deposits or dissolved in seawater.	Used in food preservation, seasoning, and de-icing roads. Sodium chloride is also used in chemical production and water softening.
Calcium Carbonate	Found in rocks such as limestone, chalk, and marble.	Used in construction as a building material and in cement production. Calcium carbonate is also used in the production of glass, paper, plastics, and as a dietary calcium supplement.
Sulfur	Sulfur is found in its elemental form in volcanic regions and in minerals like pyrite.	Used in the production of sulfuric acid, which is essential for manufacturing fertilizers, chemicals, and petroleum refining. Sulfur is also used in vulcanization of rubber and in fungicides and insecticides.

(Source: ChemLibreTexts, Creative Commons License).

Table 3.9. *Tabular Representation of Common Minerals and Uses*

Bauxite	Aluminum, foil, and aircraft components
Borax	Antiseptic detergents, welder flux, or cleaner (discovered on arid lake bottoms)
Calcite	Drugs, toothpaste, and building materials are derived from hard water deposits found in ancient seabeds.
Copper	Items such as tubing, electric lines, and sculptures
Diamond	Cutting instruments, such as blades and saws
Feldspar	Utilization in ceramics and porcelain, specifically focusing on the coloration of granites excluding the color black.
Galena	Origin of lead
Graphite	Used in Pencils and are used as a lubricant in machinery.
Gypsum	Drywall, gypsum plaster
Halite	Salt
Hematite	Origin of iron
Jade	Jewelry and ornamental figurines
Limonote/ taconite	Origin of iron
Muscovite (mica)	The white or gray substance found in electric insulators

Quartz (massive type), quartz crystal	Industries such as glass manufacture, radio production, computer manufacturing, and electronic device production
Silver	Items such as jewelry, camera equipment, and electrical devices
Sulfur	Fungicides are substances that are used to eliminate bacteria. They also have the ability to vulcanize rubber and are found in coal and fuels. Additionally, they can be used as fertilizer.
Talc	Infant powder, steatite, and gymnastics for gripping bars

(*Source:* Anslyn, Creative Commons License).

Organometallic substances are organic molecules that form bonds with metal atoms, as indicated by their name (Vardeny and Beard, 2022) (Table 3.10).

Table 3.10. *Tabular data of Examples of the Varying Reactivity of Metal Cations in Inorganic Chemistry*

Metal	**Periodic Table Group**	**Periodic Table Period**	**Reaction With Air**	**Reaction With Water**	**Reaction With Steam**	**Reaction With Dilute Acid**	**Solubility in Water**
Sodium	1	3	Burns to form oxide	Violent	Violent	Violent	Readily soluble
Calcium	2	4	Burns to form oxide	Slow	Violent	Violent	Slightly soluble
Potassium	1	4	Burns to form oxide	Violent	Violent	Violent	Readily soluble
Copper	11	4	Forms oxide on heating	No	No	No	Insoluble
Gold	11	6	No	No	No	No	Insoluble
Iron	8	4	Forms oxide on heating	No	Reversible	Reaction	Insoluble
Mercury	12	6	Reversible	No	No	No	Insoluble
Silver	11	5	No	No	No	No	Insoluble
Tin	14	5	Forms oxide on heating	No	No	Weak	Insoluble
Zinc	12	4	Forms oxide on heating	No	Reaction	Reaction	Insoluble
Lead	14	6	Forms oxide on heating	No	No	Weak	Insoluble
Aluminum	13	3	Burns to form oxide	No	Reaction	Reaction	Insoluble
Magnesium	2	3	Burns to form oxide	No	Strong	Ready	Sparingly soluble

3.3.2. Types of Bonds

Chemical compounds can be classified based on the sorts of bonds they include. Ionic compounds consist of ions and are bonded together by the electrostatic interactions between the positively and negatively charged ions. Sodium chloride, sometimes known as common salt, is a well-known example of an ionic compound. Molecular compounds consist of individual molecules that are interconnected through the sharing of electrons, known as covalent bonding. Examples include water and hydrogen fluoride (Zuckerman et al., 1986).

3.3.3. Reactivity

Another categorization system is established according to reactivity, specifically the precise sorts of chemical reactions in whereby the compounds are involved. Acids are chemicals that, when dissolved in water, generate protons (H+ ions) to form aqueous solutions (Heeb et al., 2017). They are specifically classified as proton donors. The predominant acids include HCl, H_2SO_4, HNO_3, and H_3PO_4. Conversely, bases function as proton acceptors. The hydroxide ion (OH) is the most prevalent base, as it combines with a proton (H+) to produce a water molecule (Casals et al., 2012):

$$H^+ + OH^- \rightarrow HOH\left(\textit{usually written as } H_2O\right)$$

Another category of inorganic chemicals involves substances that participate in oxidation-reduction (redox) reactions. Oxidation involves the loss of electrons, while reduction involves the gain of electrons (Braunstein, 2006). For example, in the reaction between sodium metal and chlorine gas to produce sodium chloride, electrons (e^-) are transferred from sodium atoms to chlorine atoms. This electron transfer results in the formation of sodium ions (Na^+) and chloride ions (Cl^-) in the resulting compound, sodium chloride (NaCl) (Pearson, 1988).

$$2Na + Cl_2 \rightarrow 2NaCl$$

In this chemical process, each sodium atom undergoes oxidation by losing an electron, while each chlorine atom undergoes reduction by gaining an electron. Sodium acts as the reducing agent, providing the electrons during the reaction, while chlorine acts as the oxidizing agent, accepting these electrons (Doctorovich and Di Salvo, 2007).

Metals are frequently used as reducing agents because they tend to readily lose electrons when reacting with nonmetals. Conversely, halogens such as fluorine, chlorine, and bromine, along with specific oxyanions like the permanganate ion (MnO_4^-) and the dichromate ion ($Cr_2O_7^{2-}$), are commonly encountered as oxidizing agents. These substances have a strong affinity for electrons and facilitate oxidation reactions by accepting electrons from other substances (Schaller et al., 2016).

3.3.4. Valence

Valence is a crucial notion in chemistry that is employed to ascertain the legitimacy of a chemical reaction (Jones, 1945). While the valence number of a free atom is zero, it assumes a positive or negative value in a chemical complex. When carbon undergoes combustion, the carbon atom exhibits a valence of zero, while the valence of CO_2 is +4 (Griggs and Warner, 1926):

$$C + O_2 \rightarrow CO_2$$

Each oxygen atom in CO_2 has a valence of 2, resulting in a net valence of zero for CO_2. This is a necessary condition for the chemical reaction to occur. An oxidation reaction is defined as the transition of the valence number of the carbon atom in CO_2 from 0 to +4. Another instance is the conversion of copper oxide into metallic copper through reduction (Liu et al., 2016):

$$CuO + H_2 \rightarrow Cu + H_2O$$

Furthermore, the principle guiding the classification of inorganic reactions stipulates that a substance is oxidized if its oxidation number increases (Stuyver et al., 2019).

3.3.5. Equilibrium Constant

The equilibrium constant (K) in a chemical reaction expresses the ratio of the reaction quotient (Q) at equilibrium (Ramette and Fan, 1983). The reaction quotient, Q, is determined by dividing the concentrations of reactants, each raised to their respective stoichiometric coefficients, by the concentrations of products, similarly raised to their respective stoichiometric coefficients (Sue et al., 2002).

In a random reaction,

- A+B C+D
- K ¼½ C ½ D =½ A ½ B

The symbol [A] typically denotes the concentration of substance A in a solution. Alternatively, equilibrium constants can also be expressed in terms of partial pressures when dealing with gases. These constants are determined by establishing equilibrium in a reaction, measuring the concentrations (or partial pressures) of reactants and products, and then applying these values in the appropriate equilibrium equation (Hendriks and Heikens, 2001).

In the context of acid-base reactions, the dissociation constant of an acid, denoted as Ka, is often represented by its pKa value. The pKa is defined as the negative logarithm (base 10) of Ka, expressed in units of mol/dm^3 (Harris et al., 1994). Likewise, a similar relationship exists for the dissociation constant of a base, where pKb represents the negative logarithm of Kb, the dissociation constant of the base. These values provide a standardized measure of acidity or basicity strength in aqueous solutions (Powell et al., 2004).

3.4. THE PERIODIC TABLE

The periodic table organizes chemical elements by their atomic number, electron configuration, and recurring chemical properties. It displays elements in rows and columns, highlighting periodic trends where elements with similar characteristics reside in vertical groups. Metals typically occupy the left side of each row (period), while nonmetals are found on the right (Farrell, 2020).

This table is crucial for categorizing elements based on their properties, arranged from low to high atomic number and corresponding atomic mass. Predictive patterns within the table help anticipate elemental behaviors essential across scientific fields (Figure 2.1, Tables 2.7 and 2.8) (Scerri, 2012). Each vertical column in the periodic table, known as a group or family, exhibits consistent trends in atomic radius, ionization energy, and electronegativity. Atomic radius increases down each group due to additional electron shells, reducing ionization energy as electrons become easier to remove. Electronegativity generally decreases down a group, although exceptions exist, like increasing electronegativity in group 11 (Matlin et al., 2019).

Horizontal rows, or periods, show less pronounced trends but highlight unique series like the lanthanides and actinides in the f-block. Lanthanides share physical properties and a preference for the +3 oxidation state, owing to their electron arrangement in the 4f subshell. Moving across a period from alkali metals to noble gases typically decreases atomic radius due to increased nuclear charge, resulting in higher ionization energy (Manners, 1996). Electronegativity and ionization energy increase together due to stronger electron-nucleus attraction, with electron affinity showing subtle variation across periods. Blocks on the periodic table denote elements based on their outermost electron subshell, guiding their grouping and naming conventions. The s-block encompasses groups 1 and 2, including alkali and alkaline earth metals, alongside hydrogen and helium (Silver and Phung, 2005).

Lanthanides exhibit varying degrees of hardness and resistivity, with increased paramagnetism as one moves down the series, except for ytterbium, lanthanum, and lutetium. Gadolinium displays ferromagnetism below its Curie point of 16°C (60.8°F) (Jena, 2013). Actinides share metallic traits such as a silvery appearance and susceptibility to tarnishing, despite being relatively soft. Some, like thorium, possess significant hardness comparable to mild steel when heated, facilitating their shaping into sheets and wires. Actinides are notably radioactive and paramagnetic, with electrical resistivity ranging from 15 to 150 µ-ohm cm (Yao et al., 2020).

3.5. MINERALS

A mineral is an inorganic compound naturally occurring with a crystalline structure, formed without the involvement of living organisms. Each mineral has a specific chemical composition and can exist as a single entity or as part of a rock containing multiple minerals or mineraloids (non-crystalline substances resembling minerals). Minerals are categorized based on a variety of chemical and physical properties. Differences in chemical composition and crystal structure are influenced by the geological conditions under which minerals form. Changes in pressure, temperature, or the overall composition of rock formations can alter the minerals they contain. In some jurisdictions, fossil fuels like natural gas, coal, and crude oil are classified as mineral resources, although except for oil shale, they are not considered true minerals (Hunt et al., 1950).

Key characteristics used to define minerals include:

i. Natural occurrence,
ii. Stability at normal temperatures and pressures,
iii. Definable chemical composition, and,
iv. Ordered atomic structure (Yossifova, 2007).

These criteria help distinguish minerals from other geological materials.

Nearly 99% of minerals found in the Earth's crust consist predominantly of eight elements. These elements are commonly found in various combinations with other elements to form diverse substances. Minerals are distinguished by their unique chemical compositions and crystalline structures. Over 3000 minerals have been identified, ranging from rare and precious substances like gold and diamond to more abundant minerals such as quartz (Graedel et al., 2005). Minerals can be identified by their crystalline structure and composition. They are typically found within rocks or as crushed forms like sand. Rocks are broadly categorized into three types based on their origin and formation processes (Nyfeler and Armbruster, 1998):

i. Igneous rocks,
ii. Sedimentary rocks, and
iii. Metamorphic rocks (Bandosz et al., 1992).

Igneous rocks form when molten material cools and solidifies, often from volcanic lava. Sedimentary rocks result from the accumulation and lithification of sediments, which are produced through weathering and erosion of existing rocks. Metamorphic rocks are formed when existing rocks undergo profound changes due to heat and pressure, without melting (Koutsopoulou et al., 2010).

ACTIVITY 3.1.

Explore different types of chemical bonds found in inorganic compounds. Conduct experiments or simulations to demonstrate the characteristics and strengths of ionic, covalent, and metallic bonds, illustrating their importance in material properties and reactivity.

SUMMARY

- Chapter Three of "Inorganic Chemistry" delves into the foundational aspects and classifications of inorganic compounds, offering a comprehensive exploration of their properties and applications. The chapter begins by elucidating the fundamentals, including organometallic chemistry's role and the environmental and industrial implications of inorganic chemicals. It proceeds with a detailed examination of IUPAC nomenclature, covering compounds with varying charged states, transition metals, polyatomic ions, and isotopes.
- Further, the classification of inorganic chemicals is thoroughly discussed, emphasizing their physical and chemical properties, types of bonds, reactivity patterns, and valence considerations. The chapter also explores the periodic table's significance in organizing elements and predicting their behaviors. Moreover, minerals are examined in terms of their composition, structure, and geological significance.

REVIEW QUESTIONS

1. What are the key differences between organometallic compounds and other inorganic compounds?
2. Discuss the environmental implications of inorganic chemistry with specific examples.
3. How does IUPAC nomenclature handle compounds with different charged states?
4. Explain the significance of transition metals in inorganic chemistry.
5. Compare and contrast ionic liquids with traditional ionic compounds.
6. How do polyatomic ions differ from monatomic ions in terms of structure and behavior?
7. Describe the role of equilibrium constants in understanding the reactivity of inorganic chemicals.

MULTIPLE CHOICE QUESTIONS

1. **Which organization sets the standards for chemical nomenclature, including rules for compounds with different charged states?**
   a. UNESCO
   b. IUPAC
   c. WHO
   d. EPA
2. **What term describes compounds that contain both carbon and metal atoms?**
   a. Organic compounds
   b. Inorganic compounds
   c. Organometallic compounds
   d. Ionic compounds
3. **Which of the following is NOT a type of bond commonly found in inorganic compounds?**
   a. Ionic bond
   b. Covalent bond
   c. Metallic bond
   d. Van der Waals bond
4. **What is the primary role of the periodic table in inorganic chemistry?**
   a. Predicting atomic mass
   b. Classifying elements based on atomic number
   c. Determining the reactivity of compounds
   d. Identifying isotopes
5. **Which type of ion has the same number of protons but differs in the number of neutrons?**
   a. Isotope
   b. Isobar
   c. Isotone
   d. Isomer
6. **Which property of inorganic compounds determines their ability to form stable solutions with other substances?**
   a. Equilibrium constant
   b. Reactivity
   c. Solubility
   d. Valence

7. What characteristic distinguishes minerals from other inorganic compounds?

a. They are all solid at room temperature.

b. They are found exclusively in nature.

c. They have a high molecular weight.

d. They are all highly reactive.

Answers to Multiple Questions

1. (b) 2. (c) 3. (d) 4. (b) 5. (c)

6. (a) 7. (b)

REFERENCES

1. Akingbemi, B. T., & Hardy, M. P. (2001). Oestrogenic and antiandrogenic chemicals in the environment: Effects on male reproductive health. *Annals of Medicine, 33*(6), 391–403.
2. Alemán, J. V., Chadwick, A. V., He, J., Hess, M., Horie, K., Jones, R. G., & Stepto, R. F. T. (2007). Definitions of terms relating to the structure and processing of sols, gels, networks, and inorganic-organic hybrid materials (IUPAC Recommendations 2007). *Pure and Applied Chemistry, 79*(10), 1801–1829.
3. Arnot, J. A., Mackay, D., Webster, E., & Southwood, J. M. (2006). Screening level risk assessment model for chemical fate and effects in the environment. *Environmental Science & Technology, 40*(7), 2316–2323.
4. Audrieth, L. F., & Daniels, F. (1951). Physical and inorganic chemistry. *Industrial & Engineering Chemistry, 43*(2), 269–288.
5. Bai, Q., & Kohlstedt, D. L. (1993). Effects of chemical environment on the solubility and incorporation mechanism for hydrogen in olivine. *Physics and Chemistry of Minerals, 19*, 460–471.
6. Bandosz, T. J., Jagiełło, J., Amankwah, K. A. G., & Schwarz, J. A. (1992). Chemical and structural properties of clay minerals modified by inorganic and organic material. *Clay Minerals, 27*(4), 435–444.
7. Bartholomay, R. C., Williams, L. M., & Campbell, L. J. (1996). Radionuclides, stable isotopes, inorganic constituents, and organic compounds in water from selected wells and springs from the southern boundary of the Idaho National Engineering Laboratory to the Hagerman Area, Idaho, 1995 (No. DOE/ID-22130; USGS-OFR-96-496; MISC-97-00025). Geological Survey, Idaho Falls, ID (United States). (Vol. 3, pp. 54–57).
8. Benkhaya, S., M'rabet, S., & El Harfi, A. (2020). A review on classifications, recent synthesis and applications of textile dyes. *Inorganic Chemistry Communications, 115*, 107891.

9. Bertrand, B., & Casini, A. (2014). A golden future in medicinal inorganic chemistry: The promise of anticancer gold organometallic compounds. *Dalton Transactions, 43*(11), 4209–4219.

10. Boehme, C., Uddin, J., & Frenking, G. (2000). Chemical bonding in mononuclear transition metal complexes with Group 13 diyl ligands ER (E = B-Tl): Part X: Theoretical studies of inorganic compounds. *Coordination Chemistry Reviews, 197*(1), 249–276.

11. Bondu, R., Kloppmann, W., Naumenko-Dèzes, M. O., Humez, P., & Mayer, B. (2021). Potential impacts of shale gas development on inorganic groundwater chemistry: Implications for environmental baseline assessment in shallow aquifers. *Environmental Science & Technology, 55*(14), 9657–9671.

12. Braunstein, P. (2006). Bonding and organic and inorganic reactivity of metal-coordinated phosphinoenolates and related functional phosphine-derived anions. *Chemical Reviews, 106*(1), 134–159.

13. Casals, E., Gonzalez, E., & Puntes, V. F. (2012). Reactivity of inorganic nanoparticles in biological environments: Insights into nanotoxicity mechanisms. *Journal of Physics D: Applied Physics, 45*(44), 443001.

14. Clarke, S. J., Adamson, P., Herkelrath, S. J., Rutt, O. J., Parker, D. R., Pitcher, M. J., & Smura, C. F. (2008). Structures, physical properties, and chemistry of layered oxychalcogenides and oxypnictides. *Inorganic Chemistry, 47*(19), 8473–8486.

15. Connon, R. E., Geist, J., & Werner, I. (2012). Effect-based tools for monitoring and predicting the ecotoxicological effects of chemicals in the aquatic environment. *Sensors, 12*(9), 12741–12771.

16. Day, P. (1985). Organic-inorganic layer compounds: Physical properties and chemical reactions. *Philosophical Transactions of the Royal Society of London. Series A, Mathematical and Physical Sciences, 314*(1528), 145–158.

17. Dobbs, K. D., & Hehre, W. J. (1987). Molecular orbital theory of the properties of inorganic and organometallic compounds. 6. Extended basis sets for second-row transition metals. *Journal of Computational Chemistry, 8*(6), 880–893.

18. Doctorovich, F., & Di Salvo, F. (2007). Performing organic chemistry with inorganic compounds: Electrophilic reactivity of selected nitrosyl complexes. *Accounts of Chemical Research, 40*(10), 985–993.

19. Erlander, S. R. (1968). Interaction of aqueous ions: A correlation between the hydration and solubility properties of inorganic, organic, and polymeric salts having mono- or polyatomic ions. *Journal of Macromolecular Science—Chemistry, 2*(3), 623–643.

20. Ernst, R. D. (1999). Pentadienyl ligands: Their properties, potential, and contributions to inorganic and organometallic chemistry. *Comments on Inorganic Chemistry, 21*(4–6), 285–325.

21. Fadeel, B., & Garcia-Bennett, A. E. (2010). Better safe than sorry: Understanding the toxicological properties of inorganic nanoparticles manufactured for biomedical applications. *Advanced Drug Delivery Reviews, 62*(3), 362–374.

22. Farrell, N. P. (2020). Medicinal inorganic chemistry: New perspectives and targets for the periodic table. In *Advances in Inorganic Chemistry* (Vol. 75, pp. 57–86). Academic Press.

23. Feher, F. J., & Budzichowski, T. A. (1995). Silasesquioxanes as ligands in inorganic and organometallic chemistry. *Polyhedron, 14*(22), 3239–3253.

24. Fernelius, W. C., Coyle, T. D., & Powell, W. H. (1979). Nomenclature of inorganic chemistry. 12. Isotopically modified compounds. *Pure and Applied Chemistry, 51*, 1981–1994.

25. Gao, M. R., Yuan, J., & Antonietti, M. (2017). Ionic liquids and poly (ionic liquid)s for morphosynthesis of inorganic materials. *Chemistry–A European Journal, 23*(23), 5391–5403.

26. Gibson, J. K. (1997). Chemistry of actinide and lanthanide metal ions (M+) in laser ablation of dispersions of inorganic compounds in polyimide and polytetrafluoroethylene. *Journal of Vacuum Science & Technology A: Vacuum, Surfaces, and Films, 15*(4), 2107–2118.

27. Graedel, T. E., & Howard-Grenville, J. A. (2005). Inorganic minerals and chemicals. In *Greening the Industrial Facility: Perspectives, Approaches, and Tools* (Vol. 5, pp. 189–200).

28. Griggs, M. A., & Warner, M. E. (1926). Applications of modern valence theories to some inorganic chemical reactions. *Journal of Chemical Education, 3*(4), 425.

29. Harris, W. R., Chen, Y., & Wein, K. (1994). Equilibrium constants for the binding of indium (III) to human serum transferrin. *Inorganic Chemistry, 33*(22), 4991–4998.

30. Heeb, M. B., Kristiana, I., Trogolo, D., Arey, J. S., & Von Gunten, U. (2017). Formation and reactivity of inorganic and organic chloramines and bromamines during oxidative water treatment. *Water Research, 110*, 91–101.

31. Hendriks, A. J., & Heikens, A. (2001). The power of size. 2. Rate constants and equilibrium ratios for accumulation of inorganic substances related to species weight. *Environmental Toxicology and Chemistry: An International Journal, 20*(7), 1421–1437.

32. Hoffmann, R. (1981). Theoretical organometallic chemistry. *Science, 211*(4486), 995–1002.

33. Hunt, J. M., Wisherd, M. P., & Bonham, L. C. (1950). Infrared absorption spectra of minerals and other inorganic compounds. *Analytical Chemistry, 22*(12), 1478–1497.

34. Jena, P. (2013). Beyond the periodic table of elements: The role of superatoms. *The Journal of Physical Chemistry Letters, 4*(9), 1432–1442.

35. Ji, J., Qiu, J., Wai, N., Wong, F. S., & Li, Y. (2010). Influence of organic and inorganic flocculants on physical-chemical properties of biomass and membrane-fouling rate. *Water Research, 44*(5), 1627–1635.

36. Jones, E. V. (1945). Valence, a fundamental concept in inorganic chemistry. *Journal of Chemical Education, 22*(2), 74.

37. Jorissen, W. P., Bassett, H., Damiens, A., Fichter, F., & Remy, H. (1941). Rules for naming inorganic compounds: Report of the Committee of the International Union

of Chemistry for the Reform of Inorganic Chemical Nomenclature, 1940. *Journal of the American Chemical Society, 63*(4), 889–897.

38. Josephsen, J., & Kristensen, A. K. (2006). Simulation of laboratory assignments to support students' learning of introductory inorganic chemistry. *Chemistry Education Research and Practice*, 7(4), 266–279.

39. Judeinstein, P., & Sanchez, C. (1996). Hybrid organic–inorganic materials: A land of multidisciplinarity. *Journal of Materials Chemistry, 6*(4), 511–525.

40. Kaltsoyannis, N., & McGrady, J. E. (Eds.). (2004). *Principles and Applications of Density Functional Theory in Inorganic Chemistry* (Vol. 2, pp. 98–100). Springer Science & Business Media.

41. Kang, J. Y., Lee, H. H., & Kim, K. H. (2001). Physical and chemical properties of inorganic horticultural substrates used in Korea. In *International Symposium on Growing Media and Hydroponics* (Vol. 644, pp. 237–241).

42. Kaya, S., & Kaya, C. (2015). A simple method for the calculation of lattice energies of inorganic ionic crystals based on the chemical hardness. *Inorganic Chemistry, 54*(17), 8207–8213.

43. Kirchhübel, N., & Fantke, P. (2019). Getting the chemicals right: Toward characterizing toxicity and ecotoxicity impacts of inorganic substances. *Journal of Cleaner Production, 227*, 554–565.

44. Kosmulski, M. (2001). *Chemical Properties of Material Surfaces* (Vol. 102, pp. 71–73). CRC Press.

45. Koutsopoulou, E., Papoulis, D., Tsolis-Katagas, P., & Kornaros, M. (2010). Clay minerals used in sanitary landfills for the retention of organic and inorganic pollutants. *Applied Clay Science, 49*(4), 372–382.

46. Krivovichev, S. V., Tananaev, I. G., & Myasoedov, B. F. (2007). Charge-density matching in organic–inorganic uranyl compounds. *Comptes Rendus Chimie, 10*(10, 11), 897–904.

47. Kumar, R., Kumar, R., & Prakash, O. (2019). Chapter 5: The impact of chemical fertilizers on our environment and ecosystem. In Chief Ed. *Inorganic Chemistry, 35*(69), 1173–1189.

48. Lecomte, C., Aubert, E., Legrand, V., Porcher, F., Pillet, S., Guillot, B., & Jelsch, C. (2005). Charge density research: From inorganic and molecular materials to proteins. *Zeitschrift für Kristallographie-Crystalline Materials, 220*(4), 373–384.

49. Liu, X., Wang, H., Wang, W., & Fu, Z. (2016). Simple method for the hardness estimation of inorganic crystals by the bond valence model. *Inorganic Chemistry, 55*(21), 11089–11095.

50. Mackenzie, J. D., & Bescher, E. P. (2000). Physical properties of sol-gel coatings. *Journal of Sol-Gel Science and Technology, 19*, 23–29.

51. Manners, I. (1996). Polymers and the periodic table: Recent developments in inorganic polymer science. *Angewandte Chemie International Edition in English, 35*(15), 1602–1621.

52. Matlin, S. A., Mehta, G., Hopf, H., & Krief, A. (2019). The periodic table of the

chemical elements and sustainable development. *European Journal of Inorganic Chemistry, 2019*(39, 40), 4170–4173.

53. McCleverty, J. A., & Connelly, N. G. (2001). *Nomenclature of Inorganic Chemistry II: Recommendations 2000* (Vol. 2, pp. 21–25). Royal Society of Chemistry.
54. McCusker, L. B., Liebau, F., & Engelhardt, G. (2001). Nomenclature of structural and compositional characteristics of ordered microporous and mesoporous materials with inorganic hosts (IUPAC Recommendations 2001). *Pure and Applied Chemistry, 73*(2), 381–394.
55. Mjos, K. D., & Orvig, C. (2014). Metallodrugs in medicinal inorganic chemistry. *Chemical Reviews, 114*(8), 4540–4563.
56. Mulliken, R. S. (1952). Molecular compounds and their spectra. III. The interaction of electron donors and acceptors. *The Journal of Physical Chemistry, 56*(7), 801–822.
57. Murphy, B., Murphy, C., & Hathaway, B. J. (1998). Basic principles of inorganic chemistry: Making the connections. *Royal Society of Chemistry* (Vol. 16, pp. 25–30).
58. Nyfeler, D., & Armbruster, T. (1998). Silanol groups in minerals and inorganic compounds. *American Mineralogist, 83*(1, 2), 119–125.
59. Nyquist, R. A., & Kagel, R. O. (2012). *Handbook of Infrared and Raman Spectra of Inorganic Compounds and Organic Salts: Infrared Spectra of Inorganic Compounds* (Vol. 4, pp. 21–23). Academic Press.
60. Parola, S., Julián-López, B., Carlos, L. D., & Sanchez, C. (2016). Optical properties of hybrid organic-inorganic materials and their applications. *Advanced Functional Materials, 26*(36), 6506–6544.
61. Patnaik, P. (2003). *Handbook of Inorganic Chemicals*. New York: McGraw-Hill (Vol. 529, pp. 769–771).
62. Pearson, R. G. (1988). Absolute electronegativity and hardness: Application to inorganic chemistry. *Inorganic Chemistry, 27*(4), 734–740.
63. Powell, K. J., Brown, P. L., Byrne, R. H., Gajda, T., Hefter, G., Sjöberg, S., & Wanner, H. (2004). Chemical speciation of Hg (II) with environmental inorganic ligands. *Australian Journal of Chemistry, 57*(10), 993–1000.
64. Ramette, R. W., & Fan, G. (1983). Copper (II) chloride complex equilibrium constants. *Inorganic Chemistry, 22*(22), 3323–3326.
65. Ravinder, R., Bishnoi, S., Zaki, M., & Krishnan, N. M. (2021). Revealing the compositional control of electrical, mechanical, optical, and physical properties of inorganic glasses. *arXiv preprint arXiv:2103.12050, 5*(2), 214–218.
66. Roschkowski, H., & Simmler, W. (1980). IDC-Inorganic Chemicals Data Base. Processing and storage of information in the field of inorganic chemistry. *Journal of Chemical Information and Computer Sciences, 20*(3), 154–158.
67. Sanchez, C., Lebeau, B., Chaput, F., & Boilot, J. P. (2003). Optical properties of functional hybrid organic–inorganic nanocomposites. *Advanced Materials, 15*(23), 1969–1994.
68. Scerri, E. R. (2012). The periodic table. In *Philosophy of Chemistry* (Vol. 2, pp. 329–338). North-Holland.

69. Schaller, C. P., Graham, K. J., McIntee, E. J., Jones, T. N., & Johnson, B. J. (2016). Reactivity II: A second foundation-level course in integrated organic, inorganic, and biochemistry. *Journal of Chemical Education, 93*(8), 1383–1390.

70. Selvi, D., Santhy, P., & Dhakshinamoorthy, M. (2005). Effect of inorganics alone and in combination with farmyard manure on physical properties and productivity of Vertic Haplustepts under long-term fertilization. *Journal of the Indian Society of Soil Science, 53*(3), 302–307.

71. Sharma, N., & Singhvi, R. (2017). Effects of chemical fertilizers and pesticides on human health and environment: A review. *International Journal of Agriculture, Environment and Biotechnology, 10*(6), 675–680.

72. Silver, S., & Phung, L. T. (2005). A bacterial view of the periodic table: Genes and proteins for toxic inorganic ions. *Journal of Industrial Microbiology and Biotechnology, 32*(11, 12), 587–605.

73. Slaveykova, V. I., Li, M., Worms, I. A., & Liu, W. (2020). When environmental chemistry meets ecotoxicology: Bioavailability of inorganic nanoparticles to phytoplankton. *Chimia, 74*(3), 115.

74. Stuyver, T., Danovich, D., & Shaik, S. (2019). Insights into the trends in the acidity strength of organic and inorganic compounds: A valence-bond perspective. *The Journal of Physical Chemistry A, 123*(9), 1851–1860.

75. Su, C. (2017). Environmental implications and applications of engineered nanoscale magnetite and its hybrid nanocomposites: A review of recent literature. *Journal of Hazardous Materials, 322*, 48–84.

76. Sue, K., Adschiri, T., & Arai, K. (2002). Predictive model for equilibrium constants of aqueous inorganic species at subcritical and supercritical conditions. *Industrial & Engineering Chemistry Research, 41*(13), 3298–3306.

77. Vallarino, L. M., & Wnek, G. E. (2002). Industrial applications of inorganic chemistry: A junior-senior-level interdisciplinary course. *Journal of Chemical Education, 79*(7), 832.

78. Vardeny, Z. V., & Beard, M. C. (2022). Hybrid organic inorganic perovskites: Physical properties and applications (in 4 volumes). *World Scientific* (Vol. 18, pp. 54–56).

79. Vassilev, S. V., Vassileva, C. G., Baxter, D., & Andersen, L. K. (2009). A new approach for the combined chemical and mineral classification of the inorganic matter in coal. 2. Potential applications of the classification systems. *Fuel, 88*(2), 246–254.

80. Verdonck, F., Waeterschoot, H., Van Sprang, P., Vercaigne, I., Delbeke, K., Simons, C., & Verougstraete, V. (2017). MeClas: An online tool for hazard identification and classification of complex inorganic metal-containing materials. *Regulatory Toxicology and Pharmacology, 89*, 232–239.

81. Yao, B., Kuznetsov, V. L., Xiao, T., Slocombe, D. R., Rao, C. N. R., Hensel, F., & Edwards, P. P. (2020). Metals and non-metals in the periodic table. *Philosophical Transactions of the Royal Society A, 378*(2180), 20200213.

82. Yossifova, M. G. (2007). Mineral and inorganic chemical composition of the Pernik coal, Bulgaria. *International Journal of Coal Geology, 72*(3, 4), 268–292.

83. Zuckerman, J. J., Norman, A. D., & Atwood, J. D. (1986). *Inorganic Reactions and Methods* (Vol. 15). Deerfield Beach, FL: VCH.

CHAPTER 4

Organic Chemistry

LEARNING OBJECTIVES

After studying this chapter, you will be able to:

1. Understand the principles of chemical bonding, including electronegativity and hybridization.
2. Classify organic compounds based on their structural features and functional groups.
3. Apply the IUPAC systematic approach to nomenclature for naming organic compounds.
4. Describe the nomenclature rules for various functional groups such as alcohols, ethers, aldehydes, ketones, and esters.
5. Explain the concept of sp hybridization in carbon atoms and its implications in organic chemistry.
6. Differentiate between different types of isomerism in organic compounds.
7. Discuss the significance of stereochemistry, including chirality and optical activity, in organic molecules.

KEY TERMS FROM THIS CHAPTER

Atomic theory
Functional group
IUPAC nomenclature
Stereochemistry
Electronegativity
Isomerism
Pharmaceutical

INTRODUCTION

The study of carbon along with its compounds falls under the domain of organic chemistry. It has been discovered that carbon can create an extensive range of compounds. Organic compounds play a crucial role in various aspects of daily life, including medicine, agriculture, and everyday activities (Mathey, 2003).

According to Oparin's proposal in 1923, organic chemistry could have originated from the combination of ammonia, nitrogen, carbon dioxide, and methane during the Big Bang, resulting in the formation of amino acids. This research was confirmed in the laboratory by Miller in 1950. Ancient Romans and Egyptians utilized organic chemicals from natural resources such as colors, medicines, and poisons. However, the precise chemical composition of these substances remained unidentified (Li and Chen, 2006).

In the 16th century, researchers successfully obtained pure compounds from natural sources, thanks to the work of Scheele in 1769. Additionally, analytical methods were developed by Lavoisier in 1784 to figure out the elemental composition of these compounds.

In the early 19th century, scientists held the belief that organic chemicals occurring naturally possessed a unique "vital force" that governed their synthesis, making it seemingly impossible to replicate these compounds in a laboratory setting. Luckily, in the 19th century, Frederich Wöhler made a groundbreaking discovery. Through laboratory experimentation, he discovered that urea, a compound present in urine, might be synthesized by applying heat to ammonium cyanate. The discovery made by him revealed that the synthesis of organic compounds did not rely on the natural "vital force," opening up new possibilities for chemists to create these compounds (Bachrach, 2008).

During the mid-nineteenth century, significant progress was achieved in exploring, examining, and creating numerous novel organic compounds. The exploration of organic chemistry structures originated from a bonding theory known as valence theory, which was introduced by Kekule and Couper in 1858 (Brown et al., 2014).

The field of organic chemistry flourished and became a dynamic and fruitful scientific discipline during the 19th century. Numerous innovative reaction mechanisms, synthetic processes, analytical approaches, and structural hypotheses were developed. By the conclusion of that century, significant advancements had been made in expanding the understanding of organic chemistry to encompass the analysis of biological systems, including proteins and DNA. Monthly, much knowledge is released in electronic media, journals, books, and specifically focusing on organic and biological chemistry (Faber and Faber, 2011).

Due to the abundance of material in the field of organic chemistry, new students are expected to engage in extensive study. Students are required to acquire knowledge on organic processes, analysis, mechanism, analysis, synthesis, and biological function (Streitwieser et al., 1992).

4.1. THE CHEMICAL BOND

4.1.1. Atomic Theory

The study of electrons in atoms took off in the first decade of the 1900s and gained widespread acceptance by 1926, thanks to the significant work of Heisenberg and Schroedinger. Their mathematical answers for electronic energy levels revolutionized the field, which is now known as quantum mechanics.

The electrons orbit the nucleus for an atom within energy levels. As one moves further from the nucleus, the energy of these levels increases. In the realm of chemistry, the fascinating world of energy levels is divided into shells and subshells. These shells act as containers for the subshells or orbitals, with the capacity to accommodate the maximum number of two electrons. The energy in electrons along with orbitals can be determined through calculations based on atomic theory, resulting in specific outcomes. The first two levels of energy, or shells 1 and 2, play a crucial role in bonding within the realm of organic chemistry (Chalmers, 2005) (Figure 4.1).

Shell	s	p	d	f	Total Electrons Possible
1	1				2
2	2	3			8
3	3	3	5		18
4	1	3	5	7	32

a. **Energy level 1** The 1s orbital may accommodate the maximum number of two electrons.

b. **Energy level 2** The structure consists of a maximum of eight electrons, with two in a 2s-orbital or two within each one of the three designated 2p-orbitals. The p-orbitals exhibit a distinctive barbell-like structure, aligning themselves precisely across the x, y, and z axes. The px, py, or pz orbits are commonly referred to as this (Kedrov, 1949).

c. The third energy level can hold a maximum of eighteen electrons. The arrangement consists of a duo of electrons in a 3s orbital, with six electrons distributed among all three 3p orbitals, along with 10 electrons dispersed among the five 3d orbitals.

d. The fourth energy level can hold a maximum of 32 electrons. It consists of two electrons within the 4s-orbital, six electrons in the three 4p-orbitals, 10 electrons in the five 4d-orbitals, and 14 electrons in the seven 4f-orbitals.

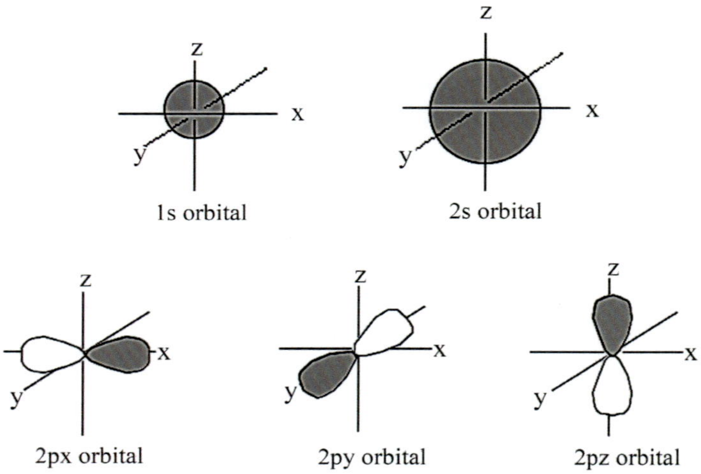

Figure 4.1. *Illustration of orbital energy shells.*

(Source: Math is love, Creative Commons License).

The Aufbau Principle states that electrons occupy the lowest energy levels before filling the higher ones. An element possesses a quantity of electrons that is equivalent to its atomic number. The electron configuration of the elements in the first and second rows is rather uncomplicated (Williams, 1927) (Table 4.1).

Table 4.1. *Tabular Representation of Elements and Their Electronic Configuration*

Element (Atomic Number)	Electron Configuration
H (1)	$1s^1$ (1st shell, s orbital, one electron)
He (2)	$1s^2$
Li (3)	$1s^2, 2s^1$
Be (4)	$1s^2, 2s^2$
B (5)	$1s^2, 2s^2, 2p^1$
C (6)	$1s^2, 2s^2, 2p^2$
N (7)	$1s^2, 2s^2, 2p^3$
O (8)	$1s^2, 2s^2, 2p^4$

F (9)	$1s^2, 2s^2, 2p^5$
Ne (10)	$1s^2, 2s^2, 2p^6$ (inert, completely filled

(*Source:* Lumen learning, Creative Commons License).

4.1.2. Electronegativity

Electronegativity is crucial in the study of chemistry. It refers to an atom's power to pull electrons towards itself. Interestingly, this power tends to rise as one moves from the left side to the right side of the periodic table.

least electronegative Li < Be < B < C < N < O < F most electronegative

Electronegativity grows in an upward direction inside a column of the periodic table.

least electronegative I < Br < Cl < F most electronegative

Electropositive elements are those that readily lose electrons and acquire a positive charge. Alkali metals exhibit a strong electropositive nature (Matsunaga et al., 2003).

4.1.3. Bonding

The bonding of atoms is affected by their electronic makeup, which decides the kind of bond that is created. Chemistry involves the formation of two main types of bonds: ionic and covalent.

4.1.3.1. Ionic Bond

The presence of ionic bonding becomes crucial when dealing with atoms that exhibit significant differences in electronegativity. In chemical bonding, one atom willingly donates an electron to another atom, which gladly accepts it. Both atoms achieve a stable noble gas configuration.

For the compound lithium fluoride, its $2s^1$ electrons of lithium are transported by the $2p^5$ orbital of fluorine. In the process, the lithium atom changes, losing an electron and transforming into

Remember

Remember that understanding the chemical bonds and structures in organic compounds is crucial for comprehending their properties and reactions.

lithium cations with üs^2 s^0 configuration. Similarly, the fluorine atom experiences a transformation, gaining an electron and becoming a fluoride anion with a different $1s^2, 2s^2, 2p^6$ configuration (Takeuchi et al., 1993).

Li	+	F	$\longrightarrow$	Li^+	+	F^-	
$1s^2$		$1s^2$		$1s^2$		$1s^2$	
$2s^1$		$2s^2$ $2p^5$		$2s^0$		$2s^2$ $2p^6$	

4.1.3.2. Covalent Bond

A covalent bond forms when two atoms share a pair of electrons, creating a strong connection between them. The hydrogen atom with one electron in its 1s orbital combines with another hydrogen atom that also has one electron in its 1s orbital. The two atoms establish a covalent link by mutually sharing a pair of electrons (Dong et al., 2015).

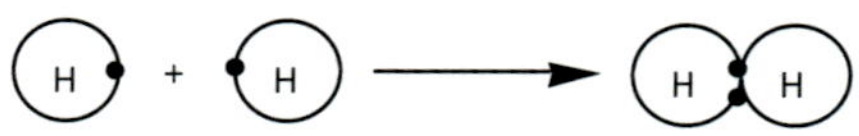

Hydrogen fluoride, HF, forms a covalent bond as hydrogen 1s electron pairs up by a $2p^5$ electrons within fluorine $(1s^2, 2s^2, 2p^5)$. This sharing of electrons allows the molecule to remain intact. In this scenario, the fluorine atom exhibits a significantly higher electronegativity compared to the hydrogen atom, causing the electrons within the bond to be more strongly attracted to the fluorine atom.

This bond is known as a covalent bond (Figure 4.2) that is where the atoms involved have a small partial charge, often represented via the Greek symbol δ (Zeng et al., 2015).

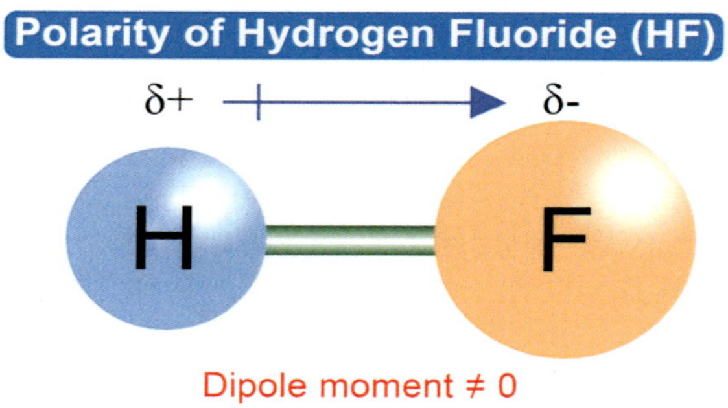

Figure 4.2. *Figure of polar covalent bond.*

(Source: Chemistry learner, Creative Commons License).

4.1.3.3. Bonding in Carbon Compounds

Carbon possesses a remarkable characteristic to establish bonds with other carbon atoms. This exceptional ability gives rise to a vast array of organic compounds, each with its own distinct set of properties. Carbon possesses the unique ability to create single, double, and triple bonds between itself and other atoms. The multiple bond capacity of carbon compounds gives them a wide range of structural possibilities. Carbon forms four bonds in all carbon compounds. Carbon atoms form two distinct kinds of bonds: sigma (σ) and pi ( ) bonds. Various combinations of these bonds result in carbon single bonds, double bonds, along triple bonds (Dyker and Bertrand, 2009).

4.1.3.4. The Carbon-Hydrogen Single Bond-The Sigma () Bond

Most bonds within carbon compounds are covalent, and a carbon-hydrogen single bond is the most common. In the carbon atom, which has an electron configuration of $\left(1s^2, 2s^2, 2p^2\right)$, $2s^2$, each of the electrons in the $2p^0$ orbital is excited to the empty $2p^0$ orbital (Frenking et al., 2014). The s or three p orbitals undergo hybridization to generate four sp3 hybrid orbitals, which possess equal energy. The electrons within the four sp^3 hybrid orbitals form bonds by overlapping with the 1s1 hydrogen orbit. A covalent link formed by the overlap of atomic orbitals is referred to as a sigma () bond. The sp^3 bonds in the molecules of methane CH_4 adopt a tetrahedral arrangement, with bond angles of 109.5o, to maximize their separation (Miller and Novoa, J2007).

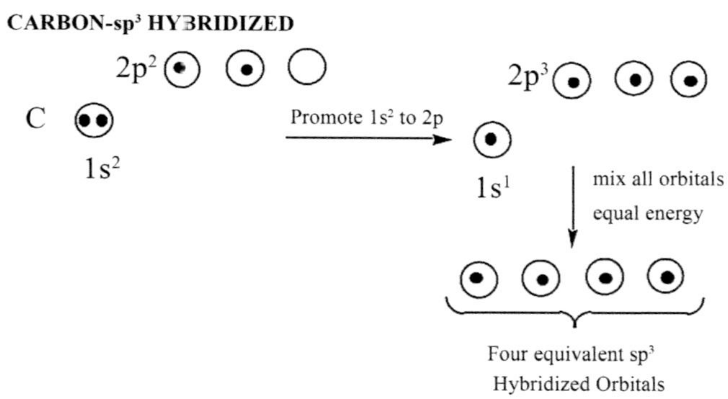

Methane, CH4

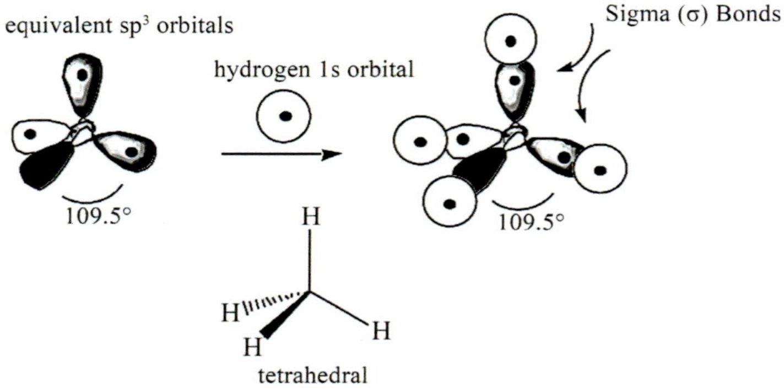

Carbon possesses a remarkable characteristic that sets it apart from other elements, enabling the creation of a vast array of diverse organic compounds: its capacity to form covalent bonds with other carbon atoms. The addition of another carbon atom to methane leads to the production of ethane. Ethane forms covalent sigma bonds by sharing hydrogen atoms and also forms the covalent sigma bond with the carbon atoms. Incorporating additional carbon atoms results in a significantly greater number of compounds (Das and Chattaraj, 2022).

Carbon-carbon sigma bond

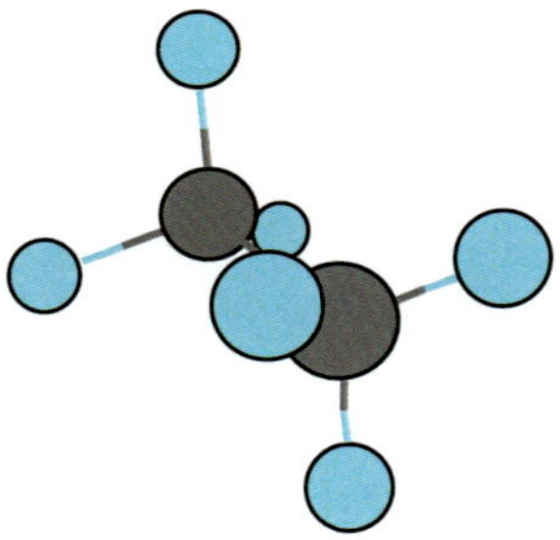

4.1.3.5. The Carbon-Carbon Double Bond-The Pi (π) Bond

Carbon forms a diverse range of compounds with carbon atoms bonded together by a double bond. These compounds are categorized as alkenes (sometimes referred to as olefins in older terminology). The orbital models below explain the double bond between carbon and carbon. The electron configuration of carbon involves the promotion of one s electron to a p orbital (Zhao et al., 2019). However, at present, just three orbitals are combined – a s orbital alongside two p orbitals. These combined orbitals, known as sp^2 hybrid orbitals, are utilized for the formation of single bonds, specifically sigma bonds. One electron is found in the p orbital (Reed and Weinhold, 1985).

When two sp^2 hybridized carbon atoms come together, they form a C-C single bond by overlapping their sp^2 orbitals. Additionally, they create another bond called a pi ( ) bond through the side-to-side overlap in their p orbitals. The molecule for ethene displayed below contains overall 5-sigma bonds along with a pi bond. The bonding within ethene leads to a planar molecule with bond angles from 120o or a C=C bond length which is longer versus the C-H bond length (Walsh, 1947).

CARBON-sp^2 HYBRIDIZED

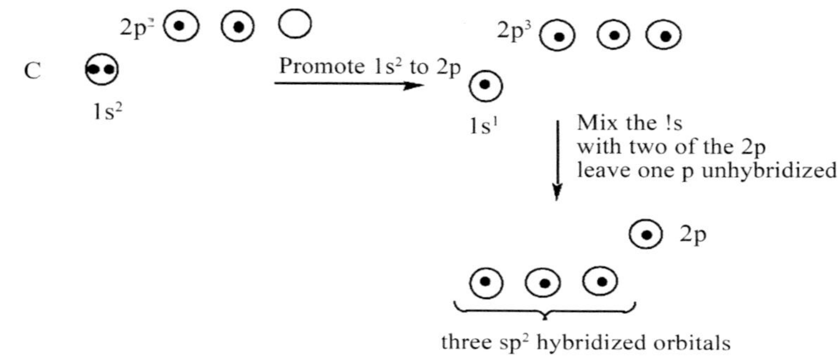

ETHENE CH_2=CH_2

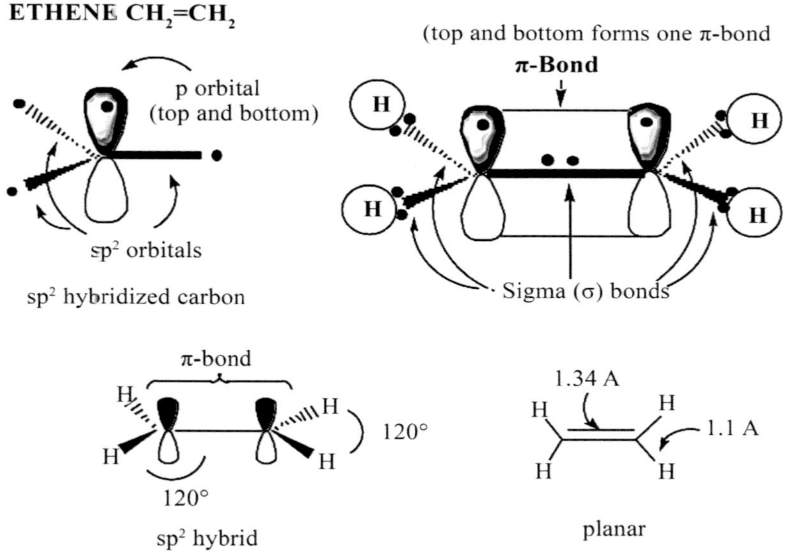

In the realm of chemistry, when carbon joins forces with an atom that is electronegative, like oxygen, nitrogen, sulfur, or a halogen, a fascinating phenomenon occurs. The resulting bond takes on a polar covalent nature, meaning that the electrons involved in the bond tend to gravitate towards the electronegative atom (Jutzi, 1975).

4.1.3.6. The Carbon-Carbon Triple Bond

Alkynes are a class of compounds that contain a triple bond, which is a different kind of bond that carbon forms with itself. After the 2s electron is promoted by the 2p orbital, its 1s orbital includes the p orbital, leading to the creation of 2 hybrid sp orbits. Two p orbitals serve to create p bonds. Thus, the carbon atom forms a sigma bond with a hydrogen atom using one of the sp hybrid orbitals, along with forming a sigma bond with another carbon atom (Putz, 2011). The transfer of hydrogen can occur between different carbon atoms through the sigma bond developed by their sp hybrid orbitals. Additionally, the two carbon atoms can be connected by two pi bonds, which result from the side-to-side overlap of their p orbitals. Ethyne molecules have a linear shape, using the hydrogen atoms positioned 180° apart. This arrangement is due to the positioning of the sp hybrid orbitals, which are also 180° apart (Frenking and Fröhlich, 2000).

4.2. CLASSIFICATION OF ORGANIC COMPOUNDS

Classification of organic compounds is commonly based on the functional groups they contain. A thorough understanding of functional groups makes it necessary for a comprehensive exploration of organic chemistry (Olsen and Nielsen, 2001).

Alkane	CH_3CH_3	Alkene	$CH_2{=}CH_2$
Alkyne	$HC{\equiv}CH$	Alkyl Halide	R-halogen
Aromatic		Alcohol	R-OH
Phenol	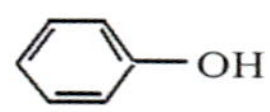	Ether	R-O-R
Aldehyde	RCH=O	Ketone	$R\text{-}\overset{O}{\overset{\parallel}{C}}\text{-}R$
Acid	RCOOH	Ester	RCOOR
Anhydride	$(RCO)_2O$	Amide	$RCONH_2$
Nitrile	RCN	Amine	$R\text{-}NH_2$

4.3. NOMENCLATURE OF ORGANIC COMPOUNDS

Given the ever-growing number of organic compounds discovered daily, coupled with the prevalence of isomers among these compounds, it becomes imperative to establish a comprehensive nomenclature system. In the realm of organic chemistry, every compound possesses its distinct molecular structure, represented by a structural formula (Styles, 1964). Furthermore, each compound requires a distinct and characteristic name to differentiate it from others. With the growth and development of organic chemistry, a multitude of compounds were assigned informal names that have become widely accepted and acknowledged. Here are a few examples:

Name	Methane	Butane	Acetone	Toluene	Acetylene	Ethyl Alcohol
Formula	CH_4	C_4H_10	CH_3COCH_3	$CH_3C_6H_5$	C_2H_2	C_2H_5OH

Common names in organic chemistry often stem from the science's historical roots and the natural sources of certain compounds. However, the connection between these names is arbitrary, lacking any rational or systematic principles for their assignment (Fletcher, 1967).

4.4. THE IUPAC SYSTEMATIC APPROACH TO NOMENCLATURE

A rational nomenclature system must achieve at least two objectives. First and foremost, it is crucial to highlight the intricate bonding patterns between carbon atoms within a specific compound, forming a distinct lattice structure of interconnected chains and rings (Cooke-Fox et al., 1989). Additionally, it is crucial to accurately identify and locate all functional groups that may be present within the compound. Given the prevalence of hydrogen in organic compounds, its quantity and positions can generally be inferred based on carbon's tetravalency, eliminating the need for explicit specification usually. The IUPAC naming system is a comprehensive set of logical guidelines developed and utilized by experts in organic chemistry to overcome challenges arising from arbitrary nomenclature. With a solid understanding of the principles and armed with a structural formula, it becomes possible to accurately assign a distinct name to each compound. Similarly, if provided with an IUPAC name, it is expected that one can accurately depict the corresponding structural formula. Typically, an IUPAC name is made up of three key components (Eller, 2006):

a. A root or base is used to indicate a significant chain or ring for carbon atoms within the molecular structure.
b. Indicators or additional components that indicate the presence of functional groups within the compound.
c. Discovering the names for substituent groups, apart from hydrogen, that contribute to the molecular structure (Moss, 2000).

Now begin by exploring compounds that lack any distinct functional groups while one delves into the IUPAC nomenclature system. These compounds consist solely of atoms of hydrogen and carbon that are bonded via sigma bonds, with all carbons being sp3 hybridized (Klinger et al., 2008).

4.4.1. Alkanes

Hydrocarbons without any triple or double bond functional group are categorized based on whether carbon atoms are organized solely in chains or also in rings. These hydrocarbons, despite lacking functional groups, serve as the foundation for functional groups found in other compound classes. They offer an excellent starting point for the study and nomenclature of organic compounds. The alkanes or cycloalkanes (Figure 4.3) are part of a broader category of compounds known as aliphatic (Atkinson, 1997). In simple terms, aliphatic compounds are somewhat compounds that do not contain any aromatic rings within their molecular structure. Here is a table that provides the IUPAC names for simple continuous-chain alkanes ranging from C-1 to C-10. These compounds are commonly identified as alkanes due to the "ane" suffix. The chain alkanes are widely recognized and can be found in numerous references along with textbooks. It is essential to commit to memory the names methane using decane, as they form the foundation for numerous IUPAC names. Thankfully, when it comes to naming chains of more than five carbon atoms, one can rely on the use of standard numerical prefixes (Schultz et al., 1990).

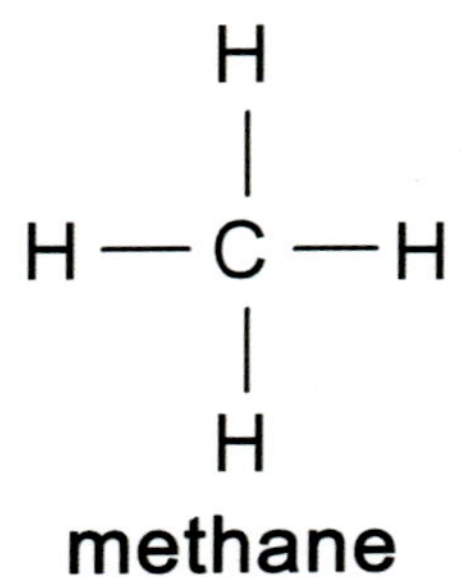

Figure 4.3. *Example of alkanes.*

(Source: Byjus, Creative Commons License).

4.4.2. Examples of Simple Unbranched Alkanes

Some examples of simple unbranched alkanes are listed in Table 4.2.

Table 4.2. *Tabular Representation of Unbranched Alkanes*

Name	**Molecular Formula**	**Structural Formula**	**Isomers**
Methane	CH_4	CH_4	1
Ethane	C_2H_6	CH_3CH_3	1
Propane	C_3H_8	$CH_3CH_2CH_3$	1
Butane	C_4H_{10}	$CH_3CH_2CH_2CH_3$	2
Pentane	C_5H_{12}	$CH_3(CH_2)_3CH_3$	3
Name	Molecular Formula	Structural Formula	Isomers
Hexane	C_6H_{14}	$CH_3(CH_2)_4CH_3$	5

Heptane	C_7H_{16}	$CH_3(CH_2)_5CH_3$	9
Octane	C_8H_{18}	$CH_3(CH_2)_6CH_3$	18
Nonane	C_9H_{20}	$CH_3(CH_2)_7CH_3$	35
Decane	$C_{10}H_{22}$	$CH_3(CH_2)_8CH_3$	75

(*Source:* Lumen learning, Creative Commons License).

Some important behavior trends and terminologies:

1. T The formulas along with structures for these alkanes exhibit a consistent increase through a CH_2 increment.
2. This type of consistent variation in several compounds is referred to as homologous.
3. All of these formulas adhere to the $\mathbf{C_nH_{2n+2}}$ rule. This represents the maximum H/C ratio achievable over a stable hydrocarbon. These compounds are considered saturated due to their maximum H/C ratio (Fokin and Schreiner, 2002).

Starting from butane (C_4H_{10}) and increasing in number as one moves to larger alkanes, one observes the presence of alkane isomers. As an illustration, there are five isomers of C_6H_{14}, represented by abbreviated line formulas (A through E) below:

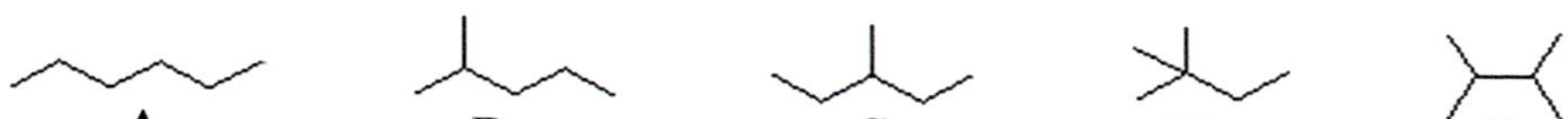

A B C D E

Despite sharing a similar molecular formula, just one of these compounds (A) can be referred to as hexane. So, what is the appropriate way to label the remaining ones?

The IUPAC system necessitates the need for names of basic, straight chains, as mentioned earlier, and also the need for names of basic alkyl groups that can be connected to these chains. Here are some typical alkyl groups provided in the table below. It's important to remember that when naming groups, the "ane" suffix can be replaced with "yl." The letter R is commonly used to represent a generic alkyl group (Gronert, 2006).

Group	CH_3–	C_2H_5–	$CH_3CH_2CH_2$–	$(CH_3)_2CH$–
Name	Methyl	Ethyl	Propyl	Isopropyl
Group	$(CH_3)_2CHCH_2$–		$CH_3CH_2CH(CH_3)$–	
Name	Isobutyl		*sec*-Butyl	

Halogen substituents can be easily incorporated into compounds using specific names for each one: chloro, fluoro, bromo, and iodo. As an illustration, the compound $(CH_3)_2CHCH_2CH_2Br$ is given the name 1-bromo-3-methylbutane. when the halogen bonds through a simple alkyl group, a different name can be used, known as an "alkyl halide." Therefore, C_2H_5Cl can be referred to as chloroethane along with ethyl chloride. Substituents like bromomethyl $BrCH_2$– and trichloromethyl CCl_3– can be included in a list and arranged alphabetically based on their complete names (Atkinson et al., 2008).

4.4.3. Cycloalkanes

Cycloalkanes consist of several carbon atom rings. The most basic examples of this group include a lone carbon ring without any substitutions, and they create a series that resembles the straight-chain alkanes. Here are the IUPAC names for the very first five members of this series, as shown in the table (Table 4.3). The final column, highlighted in yellow, provides the overall formula for a particular cycloalkane of any size. When an ordinary unbranched alkane is transformed into a cycloalkane, two hydrogen atoms must be eliminated, with one coming from both ends of the chain. Therefore, the formula of a cycloalkane with n carbon atoms is CnH_2n. Even though a cycloalkane possesses two fewer hydrogens compared to the equivalent alkane, every carbon is bonded for four other atoms, so these compounds remain believed to be fully saturated with hydrogen.

Table 4.3. *Tabular Representation of Examples of Simple Cycloalkanes*

Name	**Cyclopropane**	**Cyclobutane**	**Cyclopentane**
Molecular Formula	C_3H_6	C_4H_8	C_5H_{10}
Structural Formula	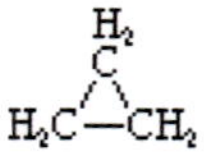	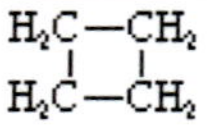	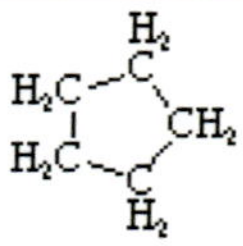
Line Formula			
Name	Cyclohexane	Cycloheptane	Cycloalkane
Molecular Formula	C_6H_{12}	C_7H_{14}	C_nH_{2n}
Structural Formula	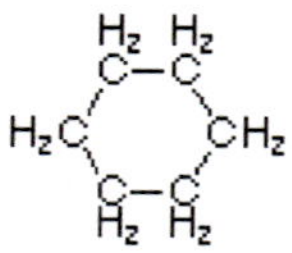	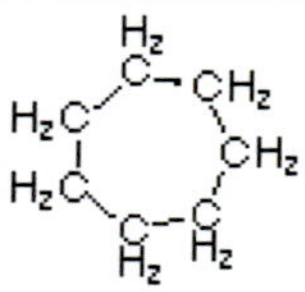	$(CH_2)_n$
Line Formula	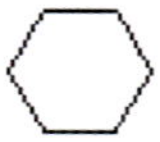		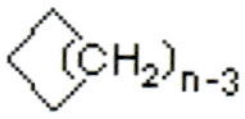

(*Source:* SIUE, Creative Commons License).

The naming convention for substituted cycloalkanes closely resembles that of branched alkanes. There is a notable distinction within laws and procedures when it comes to the numbering system. Given that a ring lacks ends like a chain, the numbering commences with the substituted ring atoms due to the equivalence of all carbons in the ring (Campbell et al., 2010).

4.4.4. Alkenes and Alkynes

Alkenes and alkynes are hydrocarbons that possess carbon-carbon double bond and carbon-carbon triple bond functional groups, respectively. The chemical formulae for these unsaturated hydrocarbons indicate the presence of numerous bonds in the functional groups (Thadathil et al., 2021):

Alkane	R–CH_2–CH_2–R	C_nH_{2n+2}	This is the highest possible hydrogen to carbon ratio for a specific quantity of carbon atoms.
Alkene	R–CH=CH–R	C_nH_{2n}	Each double bond decreases the quantity of the hydrogen atoms by 2.
Alkyne	R–C≡C–R	C_nH_{2n-2}	Each triple bond decreases the quantity of hydrogen atoms about 4.

As mentioned earlier in the Analysis for Molecular Formulas part, the molecular formula for a hydrocarbon gives insights into the potential structural types it could represent (Sato et al., 2000). Take compounds with the formula C_5H_8, for instance. Pentane, an alkane with five carbon atoms, has a molecular formula of C_5H_{12}. As a result, there is a difference of four hydrogen atoms in its composition. The observed variation indicates the possibility of compounds having a triple bond, two double bonds, one ring combined with a double bond, or two rings (Sun et al., 2020). Here are a few examples, and there have been at least 14 more!

$H_3C-C\equiv C-CH_2-CH_3$ $H_2C=CH-CH_2-CH=CH_2$

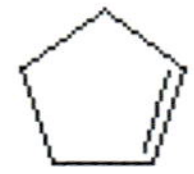

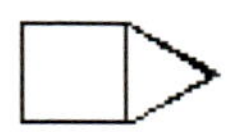

4.4.5. Benzene Derivatives

The naming conventions for compounds with substituted benzene rings are not as structured as those for alkanes, alkenes, and alkynes. Some compounds with a single substitution are named by adding a group name before "benzene," as demonstrated by the following combined names. Most of these compounds, though, are known by distinct individual names. Mastering these names requires a dedicated approach, as there is no easy substitute for memorization (Saito and Yamamoto, 2000).

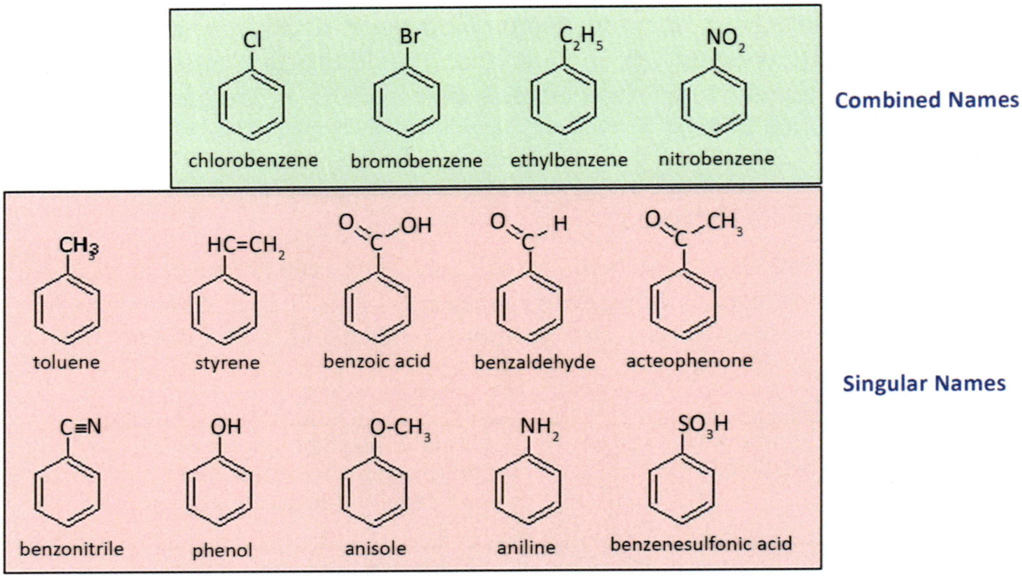

Figure 4.4. *Illustration of benzene derivatives.*

(Source: Dr. Surendran Parambadath, Creative Commons License).

Phenyl and benzyl have two substituent groups that often contain a benzene ring (Figure 4.5). They are frequently written as Ph- and Bn-, respectively. These are illustrated here in examples of their application. It's important to avoid mixing up a phenyl group using the compound phenol. A commonly used notation that can be helpful in organic chemistry is the use of "Ar-" to represent an aryl group, which refers to any aromatic ring (Krygowski et al., 2006).

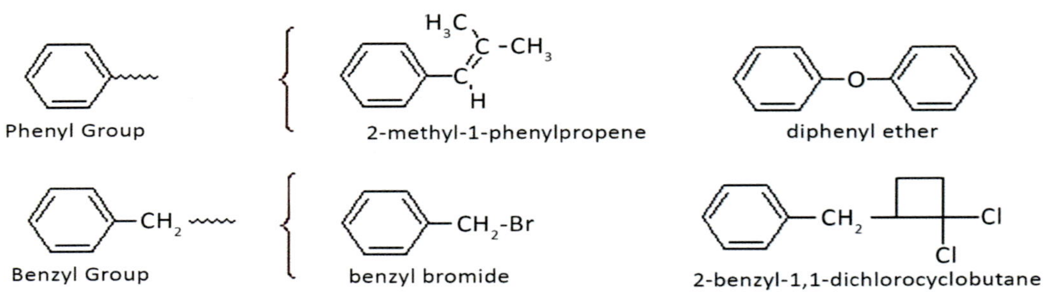

Figure 4.5. *Image of substitute groups of Phenyl and benzyl.*

(Source: Dr. Surendran Parambadath, Creative Commons License).

When multiple substituents are found upon a benzene ring, it is necessary to designate the relative positions of these substitutes by numbering the carbons upon the ring or using alternative notations. When it comes to the disubstituted bnzenes, prefixes ortho, meta, and para are frequently employed to denote a 1,2-, 1,3-, or 1,4-connection, respectively. In the next examples, the initial set of compounds demonstrate this application in red (Kaupp, 1980). Certain toluenes with two substituents have unique names (such as xylene, cresol, and toluidine), while their different arrangements are typically indicated by the prefixes ortho, meta, or para. There are specific names given to certain isomers of disubstituted benzenes, such as salicylic acid and resorcinol. Lastly, when there are at least three substituent groups, a ring is numbered in a manner that gives the substituents as few numbers as possible. This can be seen in the final row of examples. The final name arranges the substituents in alphabetical order. In the case of a symmetrical substitution (as shown in the third example starting with the left), the numbering follows the alphabetical order (Aminzai et al., 2024).

4.5. FUNCTIONAL GROUP NOMENCLATURE

Within the realm of organic chemistry, functional groups hold significant importance as it consist of a distinct set of atoms or bonds in a compound. These functional groups play a crucial role in determining the particular chemical reactions exhibited by the compound. Functional groups exhibit consistent behavior and undergo similar reactions, irrespective of the compound and are present in. Functional groups are crucial in the nomenclature of organic compounds. By merging the names of these functional groups along with the names of the parent alkanes, one can effectively differentiate between different compounds (Orozco et al., 2018).

Covalent bonds connect the atoms for a functional group to each other along with the remaining components of the compound. It is important to note the nomenclature used to describe the carbon atoms in the functional group. The carbon atom directly attached to a functional group is known as the alpha carbon, followed by the beta carbon, gamma carbon, and so on. In the realm of organic chemistry, a functional group can be classified as primary, secondary, or tertiary, based on the number of carbon atoms it is bonded to (Flynn et al., 2014).

Here are the nomenclatures for some of the other functional groups:

4.5.1. Nomenclature of Alcohols

In the IUPAC classification system, the name of alcohol is determined by removing the final "e" from the parent chain of carbons (usually an alkane, alkene, or alkyne) and adding "-ol" as the ending. When it is necessary to indicate the position of the hydroxyl group, a number is placed either between the full name of the main alkane and the suffix "-ol" (propane-1-ol) or before the name in IUPAC (1-propanol). When there's a more important group like an aldehyde, ketone, or carboxylic acid, it becomes essential to use the prefix "hydroxy-" rather than the ending "-ol."

Tip

When naming alcohols according to IUPAC rules, identify the longest carbon chain containing the hydroxyl group and assign a numerical prefix indicating its position, followed by "-ol" as the suffix.

Alcohols can be categorized into either primary or secondary groups, depending on how many carbon atoms are attached to the carbon atom carrying the hydroxyl group (Duester et al., 1999).

4.5.2. Nomenclature of Ethers

There are two methods for naming ethers. A common method involves identifying the alkyl groups adjacent to the oxygen atom within alphabetical order, followed by the term "ether." As an illustration, ethyl methyl ether is a type of ether that contains an ethyl group and a methyl group upon opposite sides of the oxygen atom. When the two groups of alkyl have the same, the ether is referred to as di[alkyl] ether. As an illustration, diethyl ether is an ether that features an ethyl group on both ends of the oxygen atom (Cooper and Purchase, 2017).

There is an alternative method for naming ethers known as the formal, IUPAC technique. Here's the structure: [short alkyl chain][oxy][long alkyl chain]. As an illustration, the name given by IUPAC over ethyl methyl ether is methoxyethane.

In cyclical ethers, the central portion of the compound is referred to as an oxacycloalkane. The presence of "oxa" in a molecule indicates that a carbon atom in the ring has been replaced by an oxygen atom. One example seems an oxacyclopentane a five-membered ring consisting of four carbon atoms and one oxygen atom.

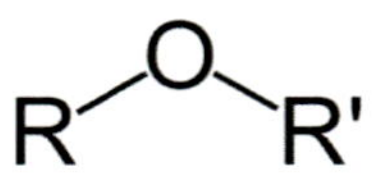

Caption: Ethers: A brief overview of the structure of the ether. Aether is defined by an oxygen atom that is bonded for two alkyl and aryl groups, as shown in this representation (Weber and Vögtle, 1980).

4.5.3. Nomenclature of Aldehydes

The IUPAC nomenclature system designates aldehydes with a distinct suffix -al. For example, the chemical formula $H_2C{=}O$ is scientifically called methanal, although it is more widely known as formaldehyde. The aldehyde-containing carbonyl group that in organic chemistry is invariably located at the terminal position of a carbon chain. Consequently, it is designated as the top position in the numerical system and does not require inclusion in the name. There are multiple simple compounds containing carbonyl groups that have names acknowledged by IUPAC (Vasiliou et al., 1999).

Furthermore, there exists a widely used approach for designating aldehydes and ketones. Aldehydes are named using common parent chain names, which are identical to those utilized by carboxylic acids. The suffix "-aldehyde" is added to the end of the name. Greek letters are frequently used to identify carbon atoms near the group of carbons in the common names of aldehydes (Figure 4.6.). The atom neighboring the carbonyl group is referred to as the alpha atom, followed by the beta atom, and so forth (Navarro-Aviño et al., 1999).

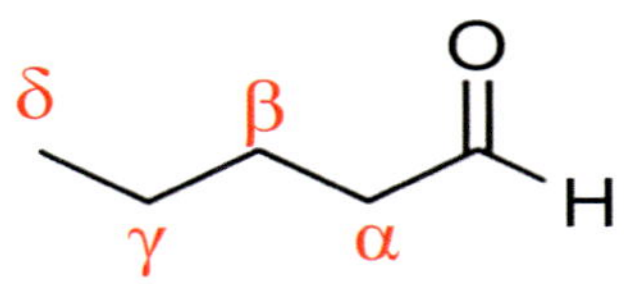

When an aldehyde moiety 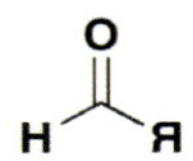 (-CHO) is connected to a ring, the name for the ring is modified by adding the suffix -carbaldehyde. In naming the ring, the carbon linked to this moiety gets assigned the #1 location number

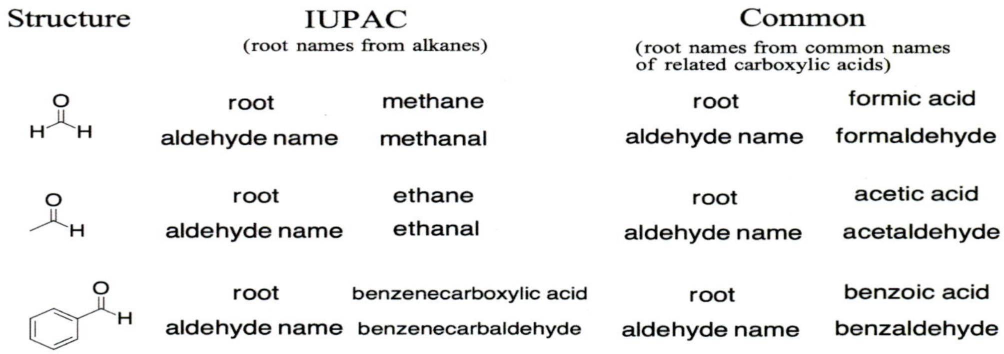

Figure 4.6. *Illustration of aldehyde common names.*

(Source: Dr. Surendran Parambadath, Creative Commons License).

4.5.4. Nomenclature of Ketones

In the worldwide system of nomenclature, a specific suffix for -one belongs to ketones. The ketones carbonyl function may be located at any position within the chain and ring, and its location is usually indicated by a number. The chain is usually numbered starting from the end nearest to the carbonyl group. Simple ketones like propanone and phenyl ethanone lack a locator number because there is just one feasible site over a ketone carbonyl function. When it comes to naming ketones, it is important to consider the alkyl groups linked to the carbonyl. By properly identifying these groups and adding the appropriate suffix, such as -ketone, one can accurately name these compounds. The alkyl groups provided are organized in alphabetical order in the given name (Neuman, 2004).

Ketone Common Names (Figure 4.7)

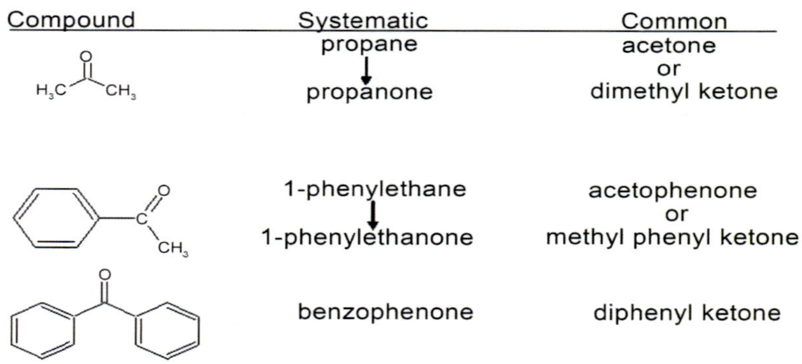

Figure 4.7. *Illustration of ketone common names.*

(Source: Dr. Surendran Parambadath, Creative Commons License).

4.5.5. Nomenclature of Ester

The term "ester" was first introduced through German chemist Leopold Gmelin in 1848, likely derived from the German word Essigäther, which translates to acetic ether.

The names of esters derive from the form of alcohol and acid they are derived from. As an illustration, the ester produced from ethanol and ethanoic acid, respectively, is commonly referred to as ethyl ethanoate. The term "ethanol" is transformed into "ethyl," while "ethanoic acid" is transformed into "ethanoate." Additional illustrations of ester names consist of methyl propanoate, derived from methanol along with propanoic acid, as well as butyl octanoate, obtained from butane or octanoic acid (Rakoff, 1975).

The compound ethyl ethanoate (Figure 4.8) gets its name from the substances used to create it: ethanol or ethanoic acid. Within this diagram, the molecule is divided into two parts: the red section, which was previously associated with ethanol (with one less hydrogen atom), and the green section, indicating the ethanoic acid-containing portion (with one less hydroxyl group). Esterification involves the removal of the H and OH elements as water, as it is a type of dehydration synthesis.

When it comes to esters derived from typical carboxylic acids, alternative, and less formal language is occasionally employed. As an illustration, ethanoic acid is commonly referred to as acetic acid, resulting in the use of "acetate" instead of "ethanoate" in the names of its esters. Alternative substitutions can be made, such as using "formate" in place of "methanoate," "propionate" rather than "propanoate," along with "butyrate" instead of "butanoate."

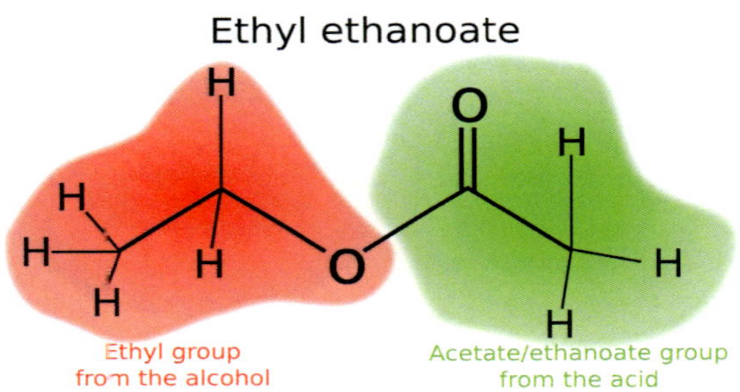

Figure 4.8. *Illustration of ethyl ethanoate.*

(Source: Geeks for Geeks, Creative Commons License).

It is common to represent the chemical formulas found in organic esters using the format RCO_2R,' where R and R' represent the hydrocarbon components of carboxylic acid or alcohol, respectively. As an illustration, butyl acetate, also known as ethanoic acid in systematic terms, is obtained via butanol and acetic acid. Its chemical formula is $CH_3CO_2C_4H_9$. Various alternative presentations are frequently used, such as BuOAc and $CH_3COOC_4H_9$. Lactones are commonly referred to as cyclic esters (Horecker et al., 1978).

4.6. CARBON--SP HYBRIDIZED

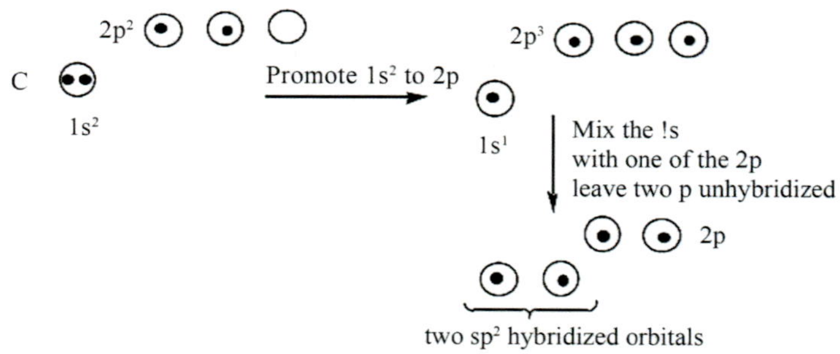

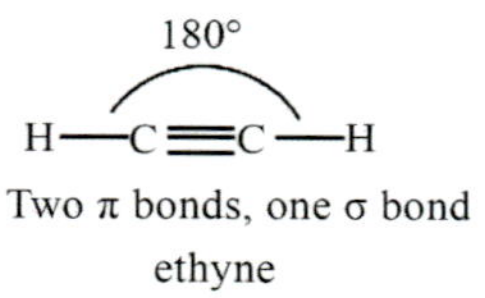

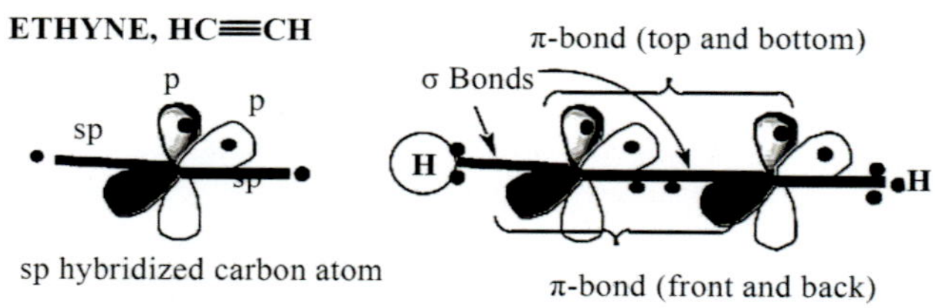

4.6.1. Polar Covalent Bonds in Carbon

Carbon can form a diverse array of chemical connections with elements other than carbon itself. The atoms participating in the bonding typically comprise oxygen, nitrogen, sulfur, and halogens. These elements have a higher electronegativity compared to carbon, which causes them to exert a stronger pull on electrons. These bonds exhibit a polarity due to the sharing of electrons. When there is a difference in charge between atoms, it leads to the presence of a dipole moment. This characteristic plays a role in determining the polarity of the molecule (Anjalikrishna et al., 2023).

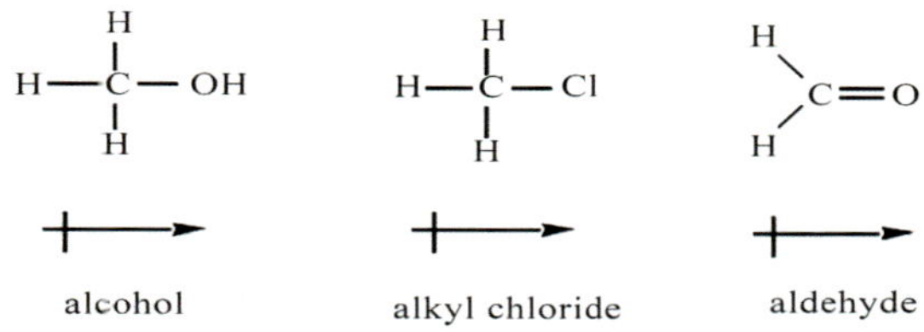

4.6.2. Hydrogen Bonds and Bond Polarity

The O-H, N-H, and F-H bonds exhibit a significant degree of polarity due to the electronegativity of the atoms involved, which causes the hydrogen atom to lose electron density. In all instances, the hydrogen atom carries a partial positive charge.

Oδ--Hδ+ Nδ--Hδ+ Fδ--Hδ+

Due to the presence of highly polarized bonds, molecules exhibit a partial positive charge on the hydrogen atom. This positive hydrogen atom is drawn towards basic sites on other molecules, including the valence electrons that are not involved in chemical interaction and are located on nitrogen and oxygen atoms. Non-bonding electrons are electrons that fulfill the octet rule of an element but do not participate in chemical bonding. Hydrogen bonding, a fascinating phenomenon, provides a valuable explanation for the unusually high boiling and melting points observed in relatively lightweight molecules. Therefore, the phenomenon of hydrogen bonding within water explains the relatively high boiling point of this three-atom compound, in comparison with various molecules of similar weight. Extra energy is required to break the bonds of hydrogen that form over the boiling process.

4.6.2.1. Hydrogen bonding in Water

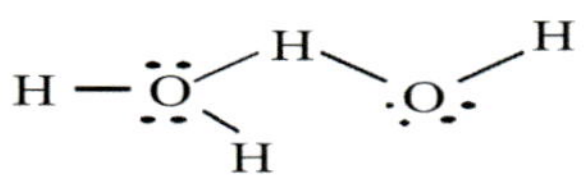

A thorough understanding of hydrogen bonding is crucial for understanding the boiling with solubility properties of alcohols and acids. This concept can be explored further in subsequent sections that delve into these particular types of molecules (Fayed et al., 2003).

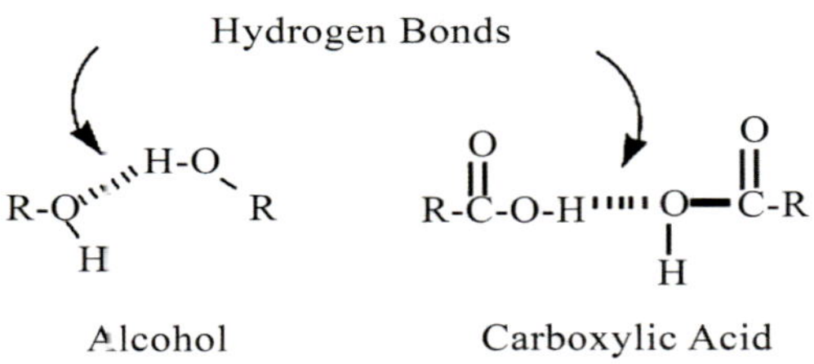

(R stands for organic function)

4.7. ORGANIC STRUCTURES

Initially, delving into the realm of organic chemistry requires grasping the concepts of energy and the intricate shapes of basic organic structures. Next, the research goes deeper into complex three-dimensional structures or ultimately investigates the analysis of reactions involving molecules derived from living organisms. Through the application of bonding principles and the analysis of molecular acidity or basicity, a deeper understanding of the broader context can be achieved. However, the ultimate objective is to enhance the well-being of humanity by advancing the creation of improved methods to be utilized in the fields of medicine, agriculture, and biomedical science, ultimately leading to a better quality of life. Progressing through the study of organic chemistry, one will become more comfortable with drawing various structures. With practice, the process of writing these structures will become easier. However, at the start, it can be quite perplexing. It's crucial to keep in mind that carbon always forms four bonds! When writing about an alkane like pentane, which consists of five carbon atoms, it is important to consider its structure and properties. One way to represent it is by illustrating carbon atoms, but this method fails to depict carbon with four bonds. It is more accurate to include all of the compound's hydrogen atoms but including every hydrogen atom over compounds can become burdensome. Therefore, one represents condensed structures as demonstrated below. Once one becomes more acquainted with it, pentane can be depicted using a line structure, at the ends along with points representing carbon atoms. The existence of hydrogens is implied without explicitly mentioning them (Etter, 1991).

C-C-C-C-C

Pentane with no H atoms

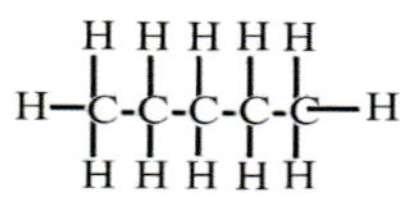

Pentane with all H atoms

CH_3-CH_2-CH_2-CH_2-CH_3

Pentane condened structure

$CH_3CH_2CH_2CH_2CH_3$

Pentane completely condensed structure

Pentane

Advanced condensed structure

There are also carbon compounds that form rings, known as cyclic compounds. Here are various representations of the compound cyclohexane. Even for the most condensed structures, it's important to acknowledge the presence of all carbon and hydrogen atoms, even if they are not explicitly written. It is essential to comprehend these abbreviated methods of representing structures to increase the speed of writing equations and improve understanding.

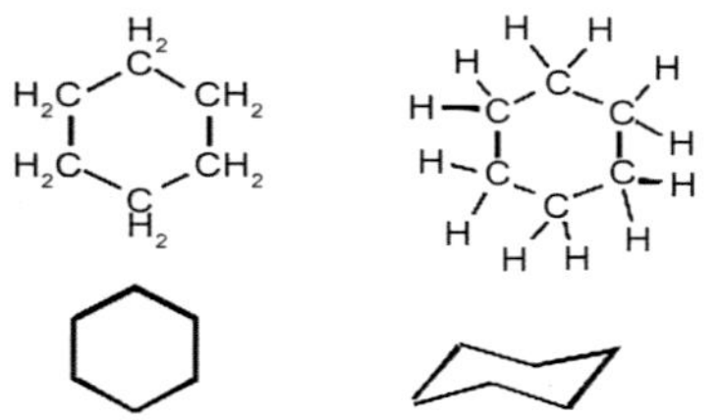

The structures of certain chemicals can be quite complex and fascinating. At times, the complexity of these structures is such that condensed forms are the most logical way to present them. As always, a carbon atom possesses four bonds (Allen and Motherwell, 2002).

Here are some complex organic structures displayed below.

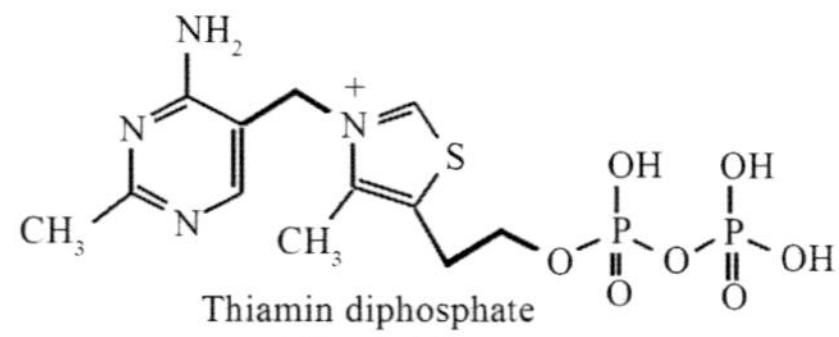

Thiamin diphosphate
Vitamin B_1
(grains, liver, pork)

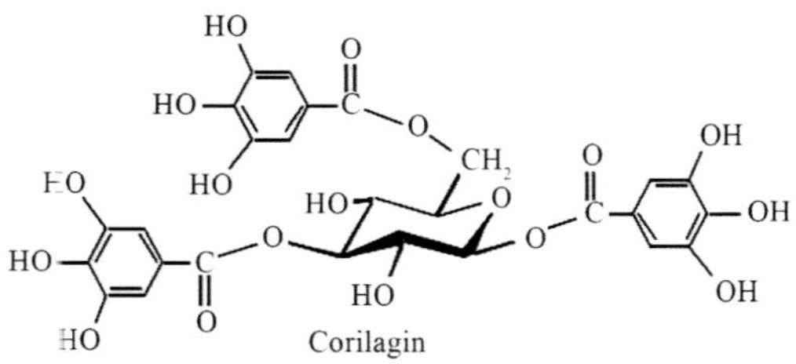

Corilagin
a molecule in the tannin family
found in cranberries
prevents bacteria from attaching to the kidneys

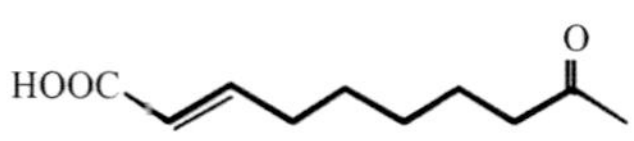

sex pheromone of bees

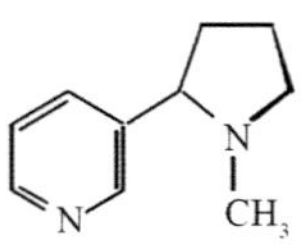

Nictotine
Addictive component
of tobacco

4.8. STEREOCHEMISTRY AND ISOMERISM

The study of stereochemistry and isomerism in chemistry explores how the arrangement of atoms for molecules affects their unique chemical and physical characteristics. Isomerism refers to compounds that have the same molecular makeup but differ in structure, while stereochemistry specifically examines the three-dimensional features of these structures. Understanding the intricate world of molecules is essential in grasping the diverse behaviors they can exhibit, especially in biological structures where the arrangement of atoms plays a vital role in determining a compound's work and effectiveness. The field of stereochemistry explores the fascinating world of chiral molecules, their fascinating relations to polarized light, along with the intriguing variations of isomerism. These variations include configurational isomers such as enantiomers and diastereomers, and conformational isomers that arise from rotations within single bonds. The knowledge obtained from stereochemistry is crucial for the development and use of molecules in various fields, including pharmaceuticals and materials science (Testa, 2013).

4.8.1. Types of Isomerism

Isomerism is an essential idea in the study of compounds, where substances using a similar molecular formula can have distinct structures or spatial configurations. Having a solid grasp on the various forms of isomerism is essential for fully comprehending the wide range and intricate nature of molecules in the field of chemistry (Makal et al., 2011).

1. **Structural Isomerism**, specifically constitutional isomerism, occurs when molecules share an identical molecular formula but have different atom connectivity. There are various subtypes of isomerism within this category:
   - i. **Chain Isomerism**: Compounds exhibit variations in the branching of their carbon chains. Take butane (C_4H_{10}) as an illustration. It can exist in the form of a straight chain (n-butane) as well as a branched chain (isobutane) (Kálmán, 2005).
   - ii. **Positional Isomerism**: This phenomenon arises when functional groups and substituents are located at various positions along a carbon chain. Consider the positioning for a hydroxyl group within 1-butanol compared to 2-butanol.

iii. **Functional Group Isomerism**: Compounds share the same molecular formula yet exhibit distinct functional groups. As an example, ethanol (C_2H_5OH) along with dimethyl ether (CH_3OCH_3) have isomers with different functional groups.

iv. **Tautomeric Isomerism**: A special form where isomers are in dynamic equilibrium, often involving the shift of a hydrogen atom and a double bond, as seen in keto-enol tautomerism.

2. **Stereoisomerism** Refers to isomers that possess identical connectivity of linked atoms but exhibit distinct three-dimensional orientations for their atoms in space. Stereoisomerism is classified into distinct categories (Hirschmann and Hanson, 1971):

   i. **Geometric Isomerism (Cis-Trans Isomerism)**: Isomerism of this nature is observed in alkenes along with cyclic compounds, where the presence of a double bond or as ring system restricts rotation and gives rise to various spatial arrangements. Take 2-butene as an example. The cis isomer has both substituents conveniently positioned upon the same side of a double bond, whereas the trans isomer places them on opposite sides.

   ii. **Optical Isomerism**: The existence of chiral centers within molecules gives rise to this phenomenon. These isomers, known as enantiomers, exhibit a unique property of being mirror images that cannot be superimposed on each other, similar to the distinction between left and right hands.

Having a solid grasp of these various forms of isomerism is crucial to accurately predicting and elucidating the characteristics and behaviors of organic compounds. It is worth noting that even though these compounds may share the same molecular formula, they frequently display distinct chemical, physical, and biological properties (Bailar, 1931).

Did you know?

Did you know that chirality in organic molecules, caused by asymmetric carbon atoms, can affect their biological activity and how they interact with other molecules?

4.8.2. Chirality and Optical Activity

Understanding chirality is crucial in the field of stereochemistry as it pertains to the unique asset of a molecule that sets it apart from its mirror image. For a molecule to be considered chiral, it must possess a chiral center, which is usually a carbon atom that is bonded to four distinct substituents. A carbon center is commonly referred to as an asymmetric along with a stereo genic center.

1. **Chirality in Organic Molecules**:
   i. **Chiral Centers**: Chirality within organic molecules often arises from a carbon atom that is bonded to four different groups. As a result, two enantiomers are produced, which are non-superimposable mirror images of each other.
   ii. **Chiral Molecules without Chiral Centers**: Certain molecules can exhibit chirality even without possessing a chiral center, thanks to their unique three-dimensional structure. Examples of such molecules include helicenes along with certain twisted or helical compounds.
2. **Optical Activity**:
   i. **Rotation of Plane-Polarized Light**: Chiral compounds possess a remarkable property of causing the rotation of polarized light. Optical activity is the term used to describe this property. When light that is polarized in a specific direction goes through the solution of a compound that has a specific arrangement of atoms, the light is either rotated to the right (referred to as "+") or shifted to the left (referred to as "−").
   ii. **Measurement and Specific Rotation**: The measurement of rotation is typically conducted through a polarimeter and is represented as the specific rotation. This standardized measure takes into consideration factors such as concentration, path length, along temperature (Balavoine et al., 1974).
3. **Enantiomers and Racemic Mixtures**:
   i. **Enantiomers**: These molecules exist as pairs that possess a unique property – they cannot be perfectly overlapped onto each other. Although they possess identical chemical and physical characteristics in a non-chiral setting, they exhibit variations in their response to polarized light and interactions with other chiral environments, such as biological systems.
   ii. **Racemic Mixtures**: When there are equal amounts of enantiomers, they combine to create a racemic mixture. In the realm of chemistry, racemic mixtures exhibit a fascinating property – they are optically inactive. This intriguing phenomenon arises from the fact that the rotations induced by each enantiomer effectively nullify one another. The concepts of chirality along with optical activity play a vital role in various industries, especially in the pharmaceutical sector. This is because the various enantiomers of a drug can exhibit significantly diverse effects within the human body. Understanding the intricacies of chiral molecules is crucial for chemical synthesis and the creation of biologically active compounds (Strauss et al., 1999).

4.8.3. Configurational Isomers

Configurational isomers tend to be a specific type of stereoisomers that can only be transformed by breaking and reestablishing covalent bonds. Enantiomers and diastereomers are essential components in the intricate world of organic molecules, each contributing significantly to their chemistry and overall functionality.

1. **Enantiomers**:
   i. **Definition and Properties**: Enantiomers are a set of molecules that exhibit a fascinating property – they are mirror images of each other but cannot be perfectly overlapped. In chiral environments and when interacting with plane-polarized light, these substances display contrasting behavior, rotating the light within opposite directions. Otherwise, their chemical and physical characteristics remain indistinguishable.
   ii. **Biological Significance**: Enantiomers can exhibit significantly varied effects within biological systems. As an example, one enantiomer for a drug can have therapeutic effects, while another may have no effect or even be harmful.
   iii. **Resolution and Separation**: Methods for isolating enantiomers, referred to as chiral resolution, play a vital role in the creation of highly pure enantiomeric substances for pharmaceutical applications. Various methods are employed in this field, such as the utilization of chiral resolving agents, asymmetric synthesis, and chiral chromatography (Gasparrini et al., 1995).
2. **Diastereomers**:
   i. **Definition and Differences**: Diastereomers are a type of stereoisomers that exhibit distinct structural differences, setting them apart from mirror image isomers. In contrast to enantiomers, diastereomers exhibit variations in their physical and chemical characteristics, including differences in melting points, boiling points, and reactivities.
   ii. **Types of Diastereomers**: Some examples consist of cis-trans isomers for alkenes along with cyclic compounds, as well as molecules with several chiral centers where certain, yet not all else, chiral centers possess opposite configurations.
   iii. **Significance in Synthesis**: Separating diastereomers is typically more straightforward compared to enantiomers because they possess distinct physical characteristics. These intermediates play a crucial role in the synthesis of intricate chiral molecules (Ramberg, 2000).

Understanding configurational isomers is crucial for comprehending the three-dimensional characteristics of organic molecules and how they behave in various chemical and biological scenarios.

4.8.4. Conformational Isomerism

Conformational isomerism occurs due to the rotation within single (sigma) bonds, enabling molecules to assume various spatial arrangements referred to as conformers. contrary to configurational isomers, conformers can interconvert without the need to break any covalent bonds.

1. **Conformations of Alkanes**:
   i. **Ethane**: An elementary illustration is when a carbon-carbon single bond rotation produces distinct conformations. The staggered or eclipsed conformations are the most notable. The staggered conformation ensures maximum separation between the hydrogen atoms on neighboring carbons, lowering repulsion and hence increasing stability. The eclipsed conformation, characterized by the alignment of hydrogen atoms, exhibits reduced stability as a result of heightened repulsion.
   ii. **Butane**: The compound displays intricate structural behavior, characterized by four primary conformers: anti (the most stable), gauche surpassed, and completely eclipsed (the least stable). The anti-conformation, characterized by the arrangement of methyl groups in an opposing orientation, reduces steric hindrance to a minimum.
2. **Conformational Analysis**:
   i. **Energy Barriers**: Conformers exhibit variations in their theoretical energy, and the conformer having the lowest energy is more stable. The energy barriers to rotation between distinct conformers play a critical role in defining the equilibrium population of each conformer.
   ii. **Newman Projections**: These tools are utilized to visually represent various conformations by observing the bond rotation axis. This illustration aids in comprehending the spatial connections and steric clashes among substituents.
3. **Cyclic Compounds and Ring Strain**:
   i. **Cyclohexane**: One classic example showcases the chair conformation as the most stable, thanks to its ability to minimize ring strain along with torsional strain. Additional conformers such as the boat or twist-boat exhibit decreased stability as a result of chemical reactions and angle strain (Bach and Dmitrenko, 2006).
   ii. **Substituted Cyclohexanes**: The arrangement of substituents can have a profound effect on stability. When considering the chair conformation, it is observed that substituents tend to favor the equatorial position to reduce 1,3-diaxial interactions along with steric hindrance.

Understanding conformational isomerism is crucial in comprehending the ever-changing nature of molecules, as it has a significant impact on the reactivity along with the physical characteristics of organic compounds. This aspect holds great importance in comprehending the conduct of flexible particles and for formulating drugs and materials with distinct conformational preferences.

4.8.5. Fischer, Newman, and Sawhorse Projections

Chemists employ a variety of projection formulas to depict the three-dimensional structure of organic molecules upon a two-dimensional plane. Several projections in organic chemistry academic writing offer distinct viewpoints for examining molecular configurations and conformations. Examples of these projections involve Fischer, Newman, and Sawhorse.

1. **Fischer Projections**:
   i. **Usage and Conventions**: Often employed in the study of molecules that possess multiple chiral centers, particularly carbohydrates along with amino acids. In Fischer projections, the horizontal lines indicate bonds that extend out of the plane, towards the viewer, while the vertical lines indicate bonds that go behind the plane.
   ii. **Application to Chirality**: Fischer projections are a valuable tool for visualizing the complete set of chiral centers in organic chemistry. The carbon with the highest oxidation state, typically the carbonyl as well as the carboxyl group, is positioned at the top. This convention proves to be quite valuable when it comes to establishing the R/S configuration while conducting comparisons between stereoisomers.
   iii. **Limitations**: Although Fischer projections are valuable in illustrating stereochemistry in a two-dimensional format, they may not provide insights into molecular conformations along with difficulties in deciphering complex molecules (Chenard and Van Zyl, 1986).
2. **Newman Projections**:
   i. **Usage and Conventions**: This tool is particularly useful for seeing the many arrangements of atoms surrounding a single bond, particularly in alkanes with cyclic systems. A Newman projection is a visual representation that allows us to observe the arrangement of substituents that rotate around a specific bond axis.
   ii. **Application to Conformational Analysis**: Understanding the stability of various conformations, like staggered along with eclipsed conformers within ethane and butane, is essential in the field of organic chemistry. They offer a comprehensive understanding of steric relationships or torsional strain.
   iii. **Advantages**: Understanding Newman projections is essential for analyzing the dimensional organization of atoms and making predictions about stable conformations. By considering steric and electronic effects, we can gain valuable insights into the behavior of molecules (Kumi et al., 2013).
3. **Sawhorse Projections**:
   i. **Usage and Conventions**: Offer a different perspective on the carbon-carbon bond along with the substituents

connected to it. This projection provides a middle ground between the two-dimensional nature of Fischer's projections along the linear perspective of Newman's projections.

ii. **Application to Stereochemistry and Conformation**: Valuable for comprehending the stereochemistry and relative energies of various conformations. Sawhorse projections are a useful tool for visualizing the interactions between atoms in butane. They can effectively depict staggered or eclipsed conformations, allowing for a clear understanding of anti and gauche interactions.

iii. **Interpretation**: Their ability to offer a more three-dimensional perspective, along with the option to rotate along with resemble a Newman projection, provides flexibility within visualizing molecular geometry.

Every projection method possesses unique strengths and applications, rendering them essential tools for chemists to comprehend and convey the intricate three-dimensional structures for organic molecules. Understanding these projections is crucial for accurately predicting reactivity, effectively designing syntheses, and gaining a comprehensive understanding of the chemical and physical characteristics of compounds in different contexts (Tong et al., 2024).

4.9. ORGANIC COMPOUNDS AND THEIR INDUSTRIAL APPLICATIONS

Organic compounds, with their carbon-based molecular structures, play a crucial role in a wide range of industrial applications thanks to their incredibly versatile chemical properties. These compounds are crucial components in a wide range of industries, including pharmaceuticals, agrochemicals, polymers, colors, and energy production. This section delves into the various roles and uses of organic substances in important industrial sectors.

4.9.1. Pharmaceuticals

The pharmaceutical industry heavily relies on organic compounds as the key components in medications that are developed to address, heal, or mitigate various illnesses. Organic molecules possess a remarkable complexity and specificity, enabling them to interact precisely with biological targets. This unique characteristic makes them highly suitable for the development of therapeutic agents.

1. **Drug Design and Synthesis**: When it comes to the creation of novel medications, the initial step involves the discovery of organic compounds that can interact with precise biological targets, including enzymes or receptors. As an example, aspirin, a well-established and extensively utilized medication, is composed of an organic compound known as acetylsalicylic acid. It functions as both an anti-inflammatory agent and a pain reliever.
2. **Chiral Compounds in Pharmaceuticals**: A significant number of drugs exhibit chirality, resulting in the existence of two enantiomers. Typically, only one enantiomer exhibits biological activity or greater efficacy. As an illustration, the medication ibuprofen is marketed to be a racemic mixture, yet it is only the S-enantiomer that delivers the therapeutic benefit that is desired. The synthesis along with the separation of these enantiomers are crucial in the production of effective medications.

3. **Biopharmaceuticals**: Recent breakthroughs in biotechnology have paved the way for the creation of biopharmaceuticals, which are intricate organic compounds such as proteins and peptides. These remarkable substances are utilized in the treatment of various ailments, including cancer, diabetes, along with autoimmune disorders. Take insulin, for instance. This remarkable peptide hormone is employed in the treatment of diabetes and is created using the advanced technique of recombinant DNA technology (Turecek et al., 2016).

4.9.2. Agrochemicals

Organic compounds play a crucial role in the agricultural industry, as they are utilized to boost crop yield and safeguard plants against pests and diseases. Agrochemicals encompass a range of substances, such as pesticides, herbicides, and fertilizers.

1. **Pesticides**: These compounds play a crucial role in safeguarding crops by eliminating or repelling pests. As an example, organophosphates along with carbamates are commonly utilized insecticides that disrupt the nervous systems of insects. Pyrethroids, which are synthetic versions of the natural pyrethrins that exist in chrysanthemum flowers, represent a highly effective class of insecticides.
2. **Herbicides**: Herbicides are used to manage undesirable plants or weeds that can hinder the growth of crops by competing for essential nutrients and sunlight. Glyphosate is a powerful herbicide that effectively controls various weed species by targeting a key enzyme responsible for plant growth inhibition.
3. **Fungicides**: These substances are employed to safeguard crops from the destructive impact of fungal infections. Organic fungicides, like azoles and strobilurins, effectively hinder the development of fungal pathogens by interfering with the cell membranes or energy generation mechanisms (Morton and Staub, 2008).

4.9.3. Polymers and Plastics

Polymers, which are made up of repeating units of organic compounds, serve as the fundamental building blocks for the plastics industry. Their remarkable adaptability and versatility have completely transformed the field of materials science, resulting in the development of numerous practical products that we use in our daily lives.

1. **Synthetic Polymers**: These materials encompass a diverse array of substances such as polyethylene, polypropylene, along polystyrene, which find applications in packaging, automotive components, and household products. Polyethylene, a widely utilized plastic, finds its application in a variety of products including plastic bags to tap water pipes.
2. **Specialty Polymers**: These products are specifically tailored to meet the needs of specialized applications that demand distinct characteristics. Take, for instance, polytetrafluoroethylene (PTFE), which is commonly referred to as Teflon. This remarkable substance is highly sought after for its exceptional non-stick qualities and remarkable resistance to various chemicals. As a result, it finds extensive use in both cookware and industrial applications.

3. **Biodegradable Polymers**: With the growing awareness of environmental issues, there is a rising fascination with the creation of biodegradable polymers sourced from renewable resources. Polylactic acid (PLA) along with polyhydroxyalkanoates (PHAs) are bioplastics that undergo natural decomposition, thereby minimizing their environmental footprint (Doppalapudi et al., 2014).

4.9.4. Dyes and Pigments

Organic compounds are essential for creating vibrant colors in textiles, paints, and inks. They have a significant impact on the manufacturing of dyes and pigments.

1. **Textile Dyes**: Various types of organic dyes, such as azo dyes, anthraquinone dyes, along with phthalocyanine dyes, play a crucial role in the textile industry. Azo dyes are known for their vibrant colors and are widely utilized, whereas anthraquinone dyes are preferred for their excellent lightfastness and are commonly employed in premium textiles.
2. **Pigments in Paints**: Organic pigments, like phthalocyanine blue along with quinacridone red, offer vibrant color and long-lasting performance in coatings and colors. The selection of these pigments is based on their durability, vibrancy, and ability to withstand fading.
3. **Inks and Printing**: Organic pigments and dyes play a crucial role within the printing industry. The inks utilized in various printing methods, such as offset, flexographic, along with digital printing, depend on the use of organic compounds to produce high-quality prints that are vivid, sharp, and long-lasting (Gregory, 1996).

ACTIVITY 4.1.

Explore the synthesis and application of organic compounds in pharmaceuticals. Investigate how specific functional groups contribute to the effectiveness and safety of drugs in medical treatments.

SUMMARY

- Chapter Four of Organic Chemistry delves into fundamental aspects crucial for understanding organic compounds. It begins by exploring the intricate details of chemical bonding, emphasizing concepts like atomic theory, electronegativity, and the diverse types of bonds that govern organic molecules. The chapter progresses to classify organic compounds based on their structures and functional groups, providing insights into the systematic nomenclature prescribed by the IUPAC guidelines. It further elucidates the properties and nomenclature rules for various functional groups such as alcohols, ethers, aldehydes, ketones, and esters.
- Additionally, the chapter delves into the concept of sp hybridization in carbon atoms, discussing its role in forming polar covalent bonds and hydrogen bonds within organic molecules. Stereochemistry and isomerism are explored comprehensively, covering types like chirality, optical activity, configurational isomers, and conformational isomerism. Practical applications of organic chemistry in industries such as pharmaceuticals, agrochemicals, polymers, and dyes are also highlighted, illustrating the broader relevance and impact of organic compounds in everyday life.

REVIEW QUESTIONS

1. Describe the concept of electronegativity and its importance in organic chemistry.
2. Explain the systematic approach to nomenclature in organic compounds using IUPAC rules.
3. Compare and contrast the different types of isomerism encountered in organic chemistry.
4. Discuss the significance of sp hybridization in carbon atoms and its role in organic molecule structure.
5. How are hydrogen bonds different from polar covalent bonds in organic compounds?
6. Explore the applications of organic chemistry in pharmaceuticals and polymers.
7. What are the key features of chirality, and why is it important in organic molecule design?

MULTIPLE CHOICE QUESTIONS

1. Which concept in organic chemistry describes the unequal sharing of electrons between atoms in a molecule?

a. Electronegativity

b. Atomic theory

c. Bond polarity

d. Molecular geometry

2. What is the primary purpose of the IUPAC nomenclature system for organic compounds?

a. To confuse students

b. To provide standardized naming conventions

c. To create complex chemical names

d. To facilitate communication among chemists

3. Which type of isomerism is characterized by molecules having the same molecular formula but different spatial arrangements of atoms?

a. Structural isomerism

b. Stereoisomerism

c. Geometric isomerism

d. Optical isomerism

4. In organic chemistry, what does sp hybridization in carbon atoms primarily involve?

a. Formation of single bonds

b. Formation of double bonds

c. Mixing of s and p orbitals

d. Formation of triple bonds

5. What is the distinguishing feature of a chiral molecule?

a. It has a mirror plane

b. It is optically inactive

c. It cannot be superimposed on its mirror image

d. It has a high symmetry

6. Which industrial sector heavily relies on organic chemistry for the production of materials like plastics?

a. Agriculture

b. Pharmaceuticals

c. Petrochemicals

d. Textiles

Answers to Multiple Questions

1. (c) 2. (b) 3. (b) 4. (c) 5. (c)

6. (c)

REFERENCES

1. Allen, F. H., & Motherwell, W. S. (2002). Applications of the Cambridge Structural Database in organic chemistry and crystal chemistry. *Acta Crystallographica Section B: Structural Science, 58*(3), 407–422.
2. Aminzai, M. T., Azizi, N., Nural, Y., & Yabalak, E. (2024). Recent progress in the reduction of benzene derivatives and their industrial applications. *Monatshefte für Chemie-Chemical Monthly, 155*(2), 115–129.
3. Anjalikrishna, P. K., Gadre, S. R., & Suresh, C. H. (2023). Topology of electrostatic potential and electron density reveals a covalent to non-covalent carbon–carbon bond continuum. *Physical Chemistry Chemical Physics, 25*(37), 25191–25204.
4. Atkinson, R. (1997). Gas-phase tropospheric chemistry of volatile organic compounds: 1. Alkanes and alkenes. *Journal of Physical and Chemical Reference Data, 26*(2), 215–290.
5. Atkinson, R., Arey, J., & Aschmann, S. M. (2008). Atmospheric chemistry of alkanes: Review and recent developments. *Atmospheric Environment, 42*(23), 5859–5871.
6. Bach, R. D., & Dmitrenko, O. (2006). The effect of carbonyl substitution on the strain energy of small ring compounds and their six-member ring reference compounds. *Journal of the American Chemical Society, 128*(14), 4598–4611.
7. Bachrach, S. M. (2008). Computational organic chemistry. *Annual Reports Section "B" (Organic Chemistry), 104*, 394–426.
8. Bailar, J. C. (1931). The study of isomerism in courses in general chemistry. *Journal of Chemical Education, 8*(2), 310.
9. Balavoine, G., Moradpour, A., & Kagan, H. B. (1974). Preparation of chiral compounds with high optical purity by irradiation with circularly polarized light, a model reaction for the prebiotic generation of optical activity. *Journal of the American Chemical Society, 96*(16), 5152–5158.
10. Brown, W. H., Poon, T., & Poon, T. (2014). *Introduction to Organic Chemistry* (Vol. 7, pp. 3–5). Hoboken, NJ, USA: John Wiley & Sons.
11. Campbell, M. J., Johnson, J. S., Parsons, A. T., Pohlhaus, P. D., & Sanders, S. D. (2010). Complexity-building annulations of strained cycloalkanes and C=O π bonds. *The Journal of Organic Chemistry, 75*(19), 6317–6325.
12. Chalmers, A. (2005). Transforming atomic chemistry into an experimental science: The limitations of Dalton's theory. *The Rutherford Journal, 1*, 89–95.
13. Chenard, B. L., & Van Zyl, C. M. (1986). Silylstannanes in organic synthesis: Scope and limitation of palladium-catalyzed reaction with acetylenes. *The Journal of Organic Chemistry, 51*(19), 3561–3566.
14. Cooke-Fox, D. I., Kirby, G. H., & Rayner, J. D. (1989). Computer translation of IUPAC systematic organic chemical nomenclature. 1. Introduction and background to a grammar-based approach. *Journal of Chemical Information and Computer Sciences, 29*(2), 101–105.

15. Cooper, C., & Purchase, R. (2017). Glossary of miscellaneous terms and techniques used in nomenclature, including colloquial terms. In *Organic Chemist's Desk Reference* (Vol. 1, pp. 229–248). CRC Press.

16. Das, P., & Chattaraj, P. K. (2022). Structure and bonding in planar hypercoordinate carbon compounds. *Chemistry, 4*(4), 1723–1756.

17. Dong, X. Q., Zhao, Q., Li, P., Chen, C., & Zhang, X. (2015). Metalorganocatalysis: Cooperating transition-metal catalysis and organocatalysis through a covalent bond. *Organic Chemistry Frontiers, 2*(10), 1425–1431.

18. Doppalapudi, S., Jain, A., Khan, W., & Domb, A. J. (2014). Biodegradable polymers—An overview. *Polymers for Advanced Technologies, 25*(5), 427–435.

19. Duester, G., Farrés, J., Felder, M. R., Holmes, R. S., Höög, J. O., Parés, X., & Jörnvall, H. (1999). Recommended nomenclature for the vertebrate alcohol dehydrogenase gene family. *Biochemical Pharmacology, 58*(3), 389–395.

20. Dyker, C. A., & Bertrand, G. (2009). Chemical bonding: Rethinking carbon. *Nature Chemistry, 1*(4), 265.

21. Eller, G. A. (2006). Improving the quality of published chemical names with nomenclature software. *Molecules, 11*(11), 915–928.

22. Etter, M. C. (1991). Hydrogen bonds as design elements in organic chemistry. *The Journal of Physical Chemistry, 95*(12), 4601–4610.

23. Faber, K., & Faber, K. (2011). *Biotransformations in Organic Chemistry: A Textbook* (Vol. 6, pp. 76–80). Berlin: Springer.

24. Fayed, T. A., Etaiw, S. E. D. H., Landgraf, S., & Grampp, G. (2003). Excited state properties of an aza-analogue of distyrylbenzene. Solvent polarity and hydrogen-bonding effects. *Photochemical & Photobiological Sciences, 2*, 376–380.

25. Fletcher, J. H. (1967). The nomenclature of organic chemistry. *Journal of Chemical Documentation,* 7(2), 64–67.

26. Flynn, A. B., Caron, J., Laroche, J., Daviau-Duguay, M., Marcoux, C., & Richard, G. (2014). Nomenclature101.com: A free, student-driven organic chemistry nomenclature learning tool. *Journal of Chemical Education, 91*(11), 1855–1859.

27. Fokin, A. A., & Schreiner, P. R. (2002). Selective alkane transformations via radicals and radical cations: Insights into the activation step from experiment and theory. *Chemical Reviews, 102*(5), 1551–1594.

28. Frenking, G., & Fröhlich, N. (2000). The nature of the bonding in transition-metal compounds. *Chemical Reviews, 100*(2), 717–774.

29. Frenking, G., Tonner, R., Klein, S., Takagi, N., Shimizu, T., Krapp, A., & Parameswaran, P. (2014). New bonding modes of carbon and heavier group 14 atoms Si–Pb. *Chemical Society Reviews, 43*(14), 5106–5139.

30. Gasparrini, F., Lunazzi, L., Misiti, D., & Villani, C. (1995). Organic stereochemistry and conformational analysis from enantioselective chromatography and dynamic nuclear magnetic resonance measurements. *Accounts of Chemical Research, 28*(4), 163–170.

31. Gregory, P. (Ed.). (1996). *Chemistry and Technology of Printing and Imaging Systems* (Vol. 2, pp. 260). Blackie Academic & Professional.

32. Gronert, S. (2006). An alternative interpretation of the C–H bond strengths of alkanes. *The Journal of Organic Chemistry, 71*(3), 1209–1219.

33. Hirschmann, H., & Hanson, K. R. (1971). Elements of stereoisomerism and prostereoismerism. *The Journal of Organic Chemistry, 36*(22), 3293–3306.

34. Horecker, B. L., Jakoby, W. B., Karlson, P., Klyne, W., Li6becq, C., & Webb, E. C. (1978). Nomenclature of phosphorus-containing compounds of biochemical importance. *Biochemical Journal, 171*, 1–19.

35. Jutzi, P. (1975). New element-carbon (p–p) π bonds. *Angewandte Chemie International Edition in English, 14*(4), 232–245.

36. Kálmán, A. (2005). Morphotropism: Link between the isostructurality, polymorphism and (stereo) isomerism of organic crystals. *Acta Crystallographica Section B: Structural Science, 61*(5), 536–547.

37. Kaupp, G. (1980). Photochemical rearrangements and fragmentations of benzene derivatives and annelated arenes. *Angewandte Chemie International Edition in English, 19*(4), 243–275.

38. Kedrov, B. M. (1949). Dalton's atomic theory and its philosophical significance. *Philosophy and Phenomenological Research, 9*(4), 644–662.

39. Klinger, R., Kolá⊠ik, C., Fluck, J., Hofmann-Apitius, M., & Friedrich, C. M. (2008). Detection of IUPAC and IUPAC-like chemical names. *Bioinformatics, 24*(13), i268–i276.

40. Krygowski, T. M., Dobrowolski, M. A., Zborowski, K., & Cyra⊠ski, M. K. (2006). Relation between the substituent effect and aromaticity. Part II. The case of meta- and para-homodisubstituted benzene derivatives. *Journal of Physical Organic Chemistry, 19*(12), 889–895.

41. Kumi, B. C., Olimpo, J. T., Bartlett, F., & Dixon, B. L. (2013). Evaluating the effectiveness of organic chemistry textbooks in promoting representational fluency and understanding of 2D–3D diagrammatic relationships. *Chemistry Education Research and Practice, 14*(2), 177–187.

42. Li, C. J., & Chen, L. (2006). Organic chemistry in water. *Chemical Society Reviews, 35*(1), 68–82.

43. Makal, T. A., Yakovenko, A. A., & Zhou, H. C. (2011). Isomerism in metal–organic frameworks: "Framework isomers." *The Journal of Physical Chemistry Letters, 2*(14), 1682–1689.

44. Mathey, F. (2003). Phospha-organic chemistry: Panorama and perspectives. *Angewandte Chemie International Edition, 42*(14), 1578–1604.

45. Matsunaga, N., Rogers, D. W., & Zavitsas, A. A. (2003). Pauling's electronegativity equation and a new corollary accurately predict bond dissociation enthalpies and enhance current understanding of the nature of the chemical bond. *The Journal of Organic Chemistry, 68*(8), 3158–3172.

46. Miller, J. S., & Novoa, J. J. (2007). Four-center carbon–carbon bonding. *Accounts of Chemical Research, 40*(3), 189–196.

47. Morton, V., & Staub, T. (2008). A short history of fungicides. *APSnet Features, 308*, 1–12.

48. Moss, G. P. (2000). Nomenclature of lignans and neolignans (IUPAC Recommendations 2000). *Pure and Applied Chemistry, 72*(8), 1493–1523.

49. Navarro-Aviño, J. P., Prasad, R., Miralles, V. J., Benito, R. M., & Serrano, R. (1999). A proposal for nomenclature of aldehyde dehydrogenases in *Saccharomyces cerevisiae* and characterization of the stress-inducible ALD2 and ALD3 genes. *Yeast, 15*(10A), 829–842.

50. Neuman, R. (2004). Carbonyl compounds: Ketones, aldehydes, carboxylic acids. *Organic Chemistry, 5*, 3–33.

51. Olsen, E., & Nielsen, F. (2001). Predicting vapour pressures of organic compounds from their chemical structure for classification according to the VOC Directive and risk assessment in general. *Molecules, 6*(4), 370–389.

52. Orozco, C., Becvar, J. E., & Narayan, M. (2018). A toolkit for mastering organic nomenclature in general chemistry. *African Journal of Chemical Education, 8*(2), 24–33.

53. Putz, M. V. (2011). *Carbon Bonding and Structures: Advances in Physics and Chemistry* (Vol. 5, pp. 76–78). Springer Science & Business Media.

54. Rakoff, H. (1975). Nomenclature of alcohols and simple esters. *Journal of the American Oil Chemists' Society, 52*(2), A65–A66.

55. Ramberg, P. J. (2000). The death of vitalism and the birth of organic chemistry: Wohler's urea synthesis and the disciplinary identity of organic chemistry. *Ambix, 47*(3), 170–195.

56. Reed, A. E., & Weinhold, F. (1985). A theoretical model of bonding in hyperlithiated carbon compounds. *Journal of the American Chemical Society, 107*(7), 1919–1921.

57. Saito, S., & Yamamoto, Y. (2000). Recent advances in the transition-metal-catalyzed regioselective approaches to polysubstituted benzene derivatives. *Chemical Reviews, 100*(8), 2901–2916.

58. Sato, F., Urabe, H., & Okamoto, S. (2000). Synthesis of organotitanium complexes from alkenes and alkynes and their synthetic applications. *Chemical Reviews, 100*(8), 2835–2886.

59. Schultz, H. P., Schultz, E. B., & Schultz, T. P. (1990). Topological organic chemistry. 2. Graph theory, matrix determinants and eigenvalues, and topological indexes of alkanes. *Journal of Chemical Information and Computer Sciences, 30*(1), 27–29.

60. Strauss, U. T., Felfer, U., & Faber, K. (1999). Biocatalytic transformation of racemates into chiral building blocks in 100% chemical yield and 100% enantiomeric excess. *Tetrahedron: Asymmetry, 10*(1), 107–117.

61. Streitwieser, A., Heathcock, C. H., Kosower, E. M., & Corfield, P. J. (1992). *Introduction to Organic Chemistry*. New York: Macmillan.

62. Styles, D. F. (1964). The classification, nomenclature and literature of carbon compounds. In Rodd's *Chemistry of Carbon Compounds* (Vol. 1, pp. 21–33). Elsevier.

63. Sun, K., Wang, X., Li, C., Wang, H., & Li, L. (2020). Recent advances in tandem selenocyclization and tellurocyclization with alkenes and alkynes. *Organic Chemistry Frontiers, 7*(19), 3100–3119.
64. Takeuchi, K., Kitagawa, T., Miyabo, A., Hori, H., & Komatsu, K. (1993). Precise control of the formation of a covalent and an ionic bond in carbocation-carbanion combination reactions. *The Journal of Organic Chemistry, 58*(21), 5802–5810.
65. Testa, B. (2013). Organic stereochemistry. Part 3: Other stereogenic elements: Axes of chirality, planes of chirality, helicity, and (E, Z)-diastereoisomerism. *Helvetica Chimica Acta, 96*(3), 351–374.
66. Thadathil, D. A., Varghese, A., & Radhakrishnan, K. V. (2021). The renaissance of electro-organic synthesis for the difunctionalization of alkenes and alkynes: A sustainable approach. *Asian Journal of Organic Chemistry, 10*(11), 2820–2847.
67. Tong, Y., Powell, C. B., Powell, G. L., & Hall, M. B. (2024). Theoretical investigation of linear relationships between the dihedral torsion angles and diosmium bond distances in diosmium sawhorse complexes. *Inorganic Chemistry, 63*(4), 1898–1908.
68. Turecek, P. L., Bossard, M. J., Schoetens, F., & Ivens, I. A. (2016). PEGylation of biopharmaceuticals: A review of chemistry and nonclinical safety information of approved drugs. *Journal of Pharmaceutical Sciences, 105*(2), 460–475.
69. Vasiliou, V., Bairoch, A., Tipton, K. F., & Nebert, D. W. (1999). Eukaryotic aldehyde dehydrogenase (ALDH) genes: Human polymorphisms, and recommended nomenclature based on divergent evolution and chromosomal mapping. *Pharmacogenetics and Genomics, 9*(4), 421–434.
70. Walsh, A. D. (1947). The properties of bonds involving carbon. *Discussions of the Faraday Society, 2*, 18–25.
71. Weber, E., & Vögtle, F. (1980). Classification and nomenclature of coronands, cryptands, podands, and their complexes. *Inorganica Chimica Acta, 45*, 65–67.
72. Williams, R. J. (1927). Ionization and the atomic structure theory in organic chemistry. *Journal of Chemical Education, 4*(7), 867.
73. Zeng, Y., Zou, R., Luo, Z., Zhang, H., Yao, X., Ma, X., & Zhao, Y. (2015). Covalent organic frameworks formed with two types of covalent bonds based on orthogonal reactions. *Journal of the American Chemical Society, 137*(3), 1020–1023.
74. Zhao, L., Pan, S., Holzmann, N., Schwerdtfeger, P., & Frenking, G. (2019). Chemical bonding and bonding models of main-group compounds. *Chemical Reviews, 119*(14), 8781–8845.

CHAPTER

5

Chemical Kinetics and Thermodynamics

LEARNING OBJECTIVES

After studying this chapter, you will be able to:

1. Understand the concept of chemical affinity and its relevance in chemical kinetics.
2. Describe the principles of modern chemical kinetics and thermodynamics.
3. Outline the procedural steps involved in chemical reactions.
4. Explain the fundamental processes of thermodynamics and their applications.
5. Discuss the unified field theory of thermodynamics and its implications.
6. Analyze the concept of linearly independent chemical reactions and entropy production.
7. Explore the extension beyond local equilibrium in chemical systems.
8. Evaluate the driving forces behind chemical reactions and their significance.

KEY TERMS FROM THIS CHAPTER

Chemical affinity
Driving force for chemical reactions
Irreversible phenomena analysis
Modern thermodynamics
Thermodynamic processes
Chemical reaction procedure
Entropy production
Modern chemical kinetics
Symmetry in thermodynamics

INTRODUCTION

The historical perspective on the study of the phenomena and energy changes involved in chemical reactions (Figure 5.1) along with their interplay is presented (Cioc et al., 2022). The study of chemical reactions has led to the emergence of two distinct branches of science: thermodynamics and chemical kinetics. These branches focus on understanding the behavior and progression of reactions in different systems (Yu and Zhang, 2017). The combination of these two fields through a universal variational principle in thermodynamics is performed. This comprehensive theory examines the time evolution within chemical kinetic systems. The significance of symmetry to be a tool for synthesis as well as integration is highlighted (Kaganovich et al., 2012).

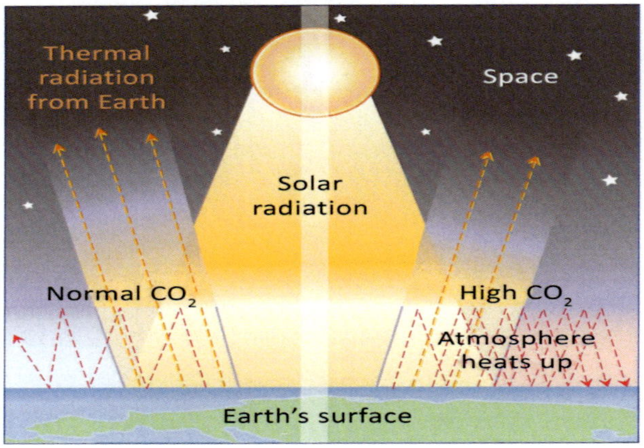

Figure 5.1. *Example of energy changes in chemical reaction.*

(Source: Saylordotorg, Creative Commons License).

5.1. CHEMICAL AFFINITY

5.1.1. Modern Chemical Kinetics

Contemporary chemical kinetics seems a field that delves into the intricate workings of chemical reactions, providing us with a comprehensive understanding of these processes. Initially, researchers explored how the rates of reactions in a system are affected by the concentrations of the components, and also investigated the impact of other factors such as electric field, temperature, radiation, and more (Christine King, 1981).

Subsequently, the focus shifted towards exploring molecular mechanisms and unique scenarios, This encompasses several processes such as the breakdown of polymers, sudden bursts of energy, the rapid burning of substances, the release of heat and light, the repetitive back-and-forth movement, reactions involving individual atoms and molecules, reactions that propagate via a series of steps, the formation of polymers, and additional phenomena (Aquilanti et al., 2018).

5.1.2. Modern Thermodynamics

Contemporary thermodynamics focuses on the study of permanent phenomena (Figure 5.2). It was built upon the principles of classical thermodynamics for equilibrium systems, often referred to as thermostatics by certain scholars. Modern thermodynamics is used to study chemical reactions that constitute non-transporting permanent procedures (Hutter, 1977).

There exist two clearly defined branches known as chemical kinetics and thermodynamics, specifically focusing on non-equilibrium or process-related phenomena. These branches have evolved into separate fields of study Phenomenological concepts of reaction rates fall into one category, and the relationship between reaction rates and thermodynamic forces belongs to another category. Now reflect on the extensive history of concepts and hypotheses surrounding chemical reactions. It may be more beneficial to disregard the terminology commonly used today and instead discuss the "transmutations of matter." It is interesting to note that Hippocrates, often referred to as the "father of medicine," believed that the driving force behind the combination of bodies is their similarity or "affinity." Despite being proven incorrect, this assumption persisted and has become a concept that has been passed down through the centuries. It now refers to the driving force behind chemical reactions (Wang, 2009).

During earlier times and up until the seventeenth century, there was a lack of experimental evidence to support theories about chemical reactions. However, in the latter part of the seventeenth century, scientific concepts regarding the factors influencing chemical reactions began to emerge. During this period within the origins of theoretical chemistry, there was a strong focus on finding solid answers to various questions: What determines whether certain substances undergo chemical reactions while others remain inert? What causes variations in the speed of different chemical reactions? And so on.

Currently, one is not inclined to delve into the intricate theories put forth to organize the available experimental data. These theories had a limited impact as the amount of experimental material quickly grew. It wasn't until the mid-nineteenth century that there was an initial effort to define the concept of affinity in terms of energy, despite its frequent usage during that time (Bejan, 2017).

The modern comprehension of the attraction over a chemical reaction, as established by Van't Hoff, a distinguished figure in the field of chemistry, originates from Gibbs' extensive thermodynamic theory concerning states of chemical equilibrium. The focus of the analysis shifted towards the driving force of the reaction, instead of the maximum work achieved by the force at hand as the reaction was advanced. However, the equation that relates the rate with the affinity remained essential (Kondepudi, 2008).

Remember

Remember that chemical kinetics and thermodynamics provide fundamental frameworks for understanding the rates of chemical reactions and the energy changes involved.

Did you know?

Did you know that modern thermodynamics extends beyond classical principles to incorporate concepts like non-equilibrium thermodynamics, which applies to systems far from equilibrium?

Figure 5.2. *Illustration of thermodynamics.*

(Source: Merchant navy decoded, Creative Commons License).

5.2. PROCEDURE OF CHEMICAL REACTION

In the realm of academia, a significant breakthrough occurred in the middle of the nineteenth century with the publication of the basic formula for chemical kinetics. This equation, a type of differential equation, provides insight into how the rate of a reaction is influenced by the concentration of its components. The subsequent solution to this equation further enhanced our understanding of chemical kinetics. The fundamental equation for chemical kinetics, known as Guldberg along with Waage's kinetic mass motion law, emerged as a result of studying the relationship between reaction rates and concentrations. This theory forms the basis for understanding how reactions proceed (Garfinkle, 2000).

Therefore, the study of the chemical process has split into two branches: (a) Phenomenological studies in chemical kinetics focus on analyzing the rates of chemical reactions and their temporal variations within chemically kinetic systems. A significant volume of experimental research was conducted, and the appropriate differential equations were formulated and solved, during a few decades near the beginning of the century. (b) In the field of chemical thermodynamics, researchers gather experimental data by measuring the changes in thermodynamic functions such as entropy, enthalpy, free energy, and chemical potentials. These measurements are taken as chemical reactions occur along a path that passes through equilibrium states. Extensive research has been conducted regarding the thermodynamics underlying chemical equilibria for a wide range of reactions. An analysis was conducted on the correlation within equilibrium constants or rate constants (Figure 5.3).

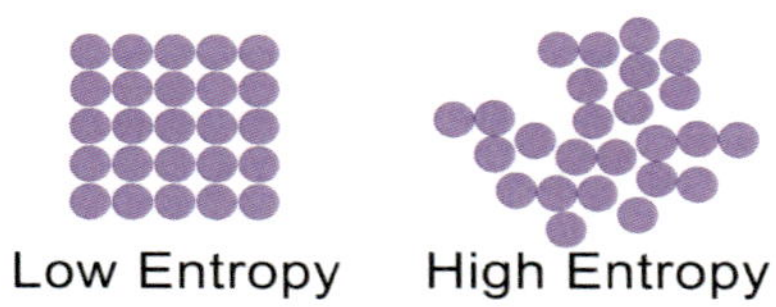

Figure 5.3. *Illustration of entropy.*

(Source: Shiken, Creative Commons License).

A significant connection was established between the realms of kinetics and the traditional theory of thermodynamics in the study of chemical reactions. An equation that aligns with the Guldberg—Waage equation for chemical kinetics was discovered through heuristic methods, specifically for ideal systems such as ideal gases along with ideal dilute solutions.

The connection between the rate at which chemical reactions occur and the principles of thermodynamics that apply to systems not in equilibrium. was discovered early on, marking an important milestone in the field's development. In the context of the 1980s, regarding a basic, reversible chemical reaction

$$\sum_{k=1}^{K} v_k^{'} B_k \rightleftarrows \sum_{k=1}^{K} v_k^{''} B_k, \qquad (5.1)$$

Among the various components B1, B2...The Guldberg-Waage kinetic equation, which represents the net reaction rate, is commonly referred to as BK.

$$\frac{d\xi}{dt} = \frac{d\vec{\xi}}{dt} - \frac{d\overleftarrow{\xi}}{dt}, \qquad (5.2)$$

This can be expressed as

$$\frac{d\xi}{dt} = \vec{k}\prod_{k=1}^{K} c_k^{v_k^{ü}} - \overleftarrow{k}\prod_{k=1}^{K} c_k^{v_k}, \qquad (5.3)$$

In elementary reactions, the stoichiometric coefficients, represented by the "vs," correspond to the numbers for colliding particles. These coefficients can be zero, one, or two.

The equation that relates the chemical potentials is crucial for understanding the process of transformation (Santillán, 2014).

$$\mu_k = \left(\frac{\partial g}{\partial c_k}\right)_{T,p,c_{k\neq j}} ; \qquad k = 1, 2, ..., K, \qquad (5.4)$$

Given the ratios (c) of the constituents in a hypothetical system

$$\mu_k = \mu_k^{\theta} + RT \ln\left(c_k / c_k^{\theta}\right); \qquad k = 1, 2, ..., K, \qquad (5.5)$$

The symbol "e" is used to represent specific typical values, while "g" is used to indicate the density of the Gibbs functions.

Through the substitution of equation (5.5) for the kinetic equation, the Marcelin—Kohnstamm equation can be derived.

$$\frac{d\xi}{dt} = \lambda\left(e^{A'/RT} - e^{A''/RT}\right) \qquad (5.6)$$

or

$$\frac{d\vec{\xi}}{dt} = \vec{\lambda} e^{A'/RT}; \quad \frac{d\bar{\xi}}{dt} = \bar{\lambda} e^{A''/RT} \tag{5.7}$$

with

$$\vec{\lambda} = \exp\left(-\sum_{k=1}^{K} v_k^{'} \mu_k^{\theta}/RT\right)\prod_{k=1}^{K}\left(c_k^{\theta}\right)^{v_k^{'}}; \bar{\lambda} = \exp\left(-\sum_{k=1}^{K} v_k^{"} \mu_k^{\theta}/RT\right)\prod_{k=1}^{K}\left(c_k^{\theta}\right)^{v_k^{"}} \tag{5.8}$$

$$A' = \sum_{k=1}^{K} v_k^{ü} \mu_k; \quad A" = \sum_{k=1}^{K} v_k \mu_k. \tag{5.9}$$

Here, it is important to note that the chemical reaction is assumed to be reversible, as indicated by the double arrow in the chemical equation (5.1). Another condition to consider is the affinity X, The advancement of the forward reaction is balanced by the backward affinity, as indicated by the symbol – representing equilibrium values. These equilibrium conditions require that I = 1. Thus, the arrows will not be included above the symbol 1.

It is important to keep in mind that the system's free energy is altered by the reaction over a specific period by a certain amount.

$$-\sum_{k=1}^{K} v_k^{'} \frac{d\xi}{dt} \mu_k + \sum_{k=1}^{K} v_k^{"} \frac{d\xi}{dt} \mu_k = -A\frac{d\xi}{dt} \tag{5.10}$$

The value is equivalent to the inverse of the maximal work capacity of the entire system. The importance

$$A = A' - A" \tag{5.11}$$

The affinity towards the reaction, as defined by van't Hoff, denotes the quantification of the tendency of a reaction to proceed toward the formation of products (Keizer, 1982).

In 1850, Clausius established the concept of "non-compensated heat," which could potentially be linked to the affinity mentioned before. In contemporary terminology, the relationship can be expressed as:

$$\frac{\partial S}{\partial \xi} = \frac{A}{T} \tag{5.12}$$

In the context of this system, we consider T and S as the temperature and entropy, respectively. Additionally, g' represents the degree of the reaction.

During that time, there was ongoing research on the experimental and theoretical aspects of chemical reactions in the 1910s and 1920s. Additionally, a connection with thermodynamics was never initially acknowledged but was discovered by incorporating the concept of the activity coefficient into the Guldberg-Waage equation for chemical reactions within electrolyte solutions. An exploration of the relationship between the activity coefficient and the Gibbs function, chemical potential, along relative activity would have revealed the connection between these concepts. Unfortunately, this did not occur, as the connection

a_k = YkQ/ck*

Another approach was employed for determining an activity coefficient Yk, which is dependent on temperature, pressure, and composition. As a result, the Guldberg-Waage equation did not accept a direct substitution of concentrations with relative activities (White, 1997).

5.3. PROCESS OF THERMODYNAMICS

After some time, the groundwork for the study of irreversible events in thermodynamics was established, albeit with a focus on linear constitutive equations. Put simply, it was demonstrated that the coefficient matrix associated with constitutive equations, which connect fluxes and forces, exhibits symmetry. Furthermore, it has been proved that reciprocal reactions may be expressed using a potential, allowing for the development of a variational principle known as the "principle of minimal dissipation of energy." The development of the constitutive equation involving heat conduction across a continuous medium was achieved using this idea (Ishida and Chuang, 1997).

Following several significant contributions, including Onsager's, a theoretical framework for understanding irreversible procedures in the continuous medium was established. Additionally, another variational principle, known as the " principle of minimum creation of entropy," was identified. In the study of irreversible procedures in continua, it has been demonstrated that these processes result in the production of entropy or the dissipation of energy. This means that the conversion of the dissipated energy to useful work becomes unattainable (Alonso and Ydstie, 1996). The rate at which entropy is produced per unit volume can be expressed as a bilinear equation, where each term represents the product from the Cartesian aspect of flux with the corresponding component of the conjugate force. In the context of scientific study, In the phenomena of viscous flow, a viscous pressure tensor is equivalent to the current density, whereas the vector gradients of the central point for mass velocity divided by the temperature reflects the tensorial thermodynamics force (Berg, 1980). It is fascinating to explore the current densities associated with the diffusion of chemical components, which can be understood as the fluxes and negative gradients of chemical potentials divided from the temperature, serving as the driving forces within diffusion processes. In the realm of chemical reactions, the current densities along with forces are expressed as scalars (Eu, 2002). The reaction rates represent the fluxes, while the affinities divided by the temperature serve as the forces. The constitutive equations allow for the interdependence of flux and force elements within the system. The equations involve specific values that remain constant throughout the analysis. Fourier's law of heat transfer exemplifies constitutive equations and material constants for heat conductivity. Understanding the fundamental equations is crucial for performing accurate calculations. The generation of entropy can be expressed in a mathematical form (Chen and Mathias, 2002).

$$\sigma = \sum_{i=1}^{f} J_i X_i, \tag{5.14}$$

where the Js represent Cartesian components of the fluxes and the Xs are the Cartesian components of the forces.

In the linear theory, from the linear constitutive equations

$$J_i \aleph \sum_{j=1}^{f} \pm_{ij} X_j, \; X_i \quad \sum_{j=1}^{f} \;_{ij} \;_j, \; i \quad 1,2,\ldots,f, \tag{5.15}$$

and the inverse relationships

$$L_{ij} = L_{ji}, \qquad R_{ij} = R_{ji}, \qquad i,j = 1,2,\ldots,f, \tag{5.16}$$

The presence of a flux potential

$$\psi = \frac{1}{2}\sum_{i,j=1}^{f} L_{ij} X_i X_j \tag{5.17}$$

along with the potential energy

$$\phi = \frac{1}{2}\sum_{i,j=1}^{f} R_{ij} J_i J_j, \tag{5.18}$$

Deriving the flux components along with force components is a logical consequence of this

$$J_i = \frac{\partial \psi}{\partial X_i}; \qquad X_i = \frac{\partial \phi}{\partial J_i}; \qquad i = 1,2,\ldots,f. \tag{5.19}$$

After nearly 35 years since Onsager's recognition of "the principle for the least dissipation for energy" and 20 years following Prigogine's inclusion of the "principle of minimum creation of entropy," a significant breakthrough was made in the field of thermodynamics. The formulation of their "governing principle for dissipative processes" marked a major advancement in our understanding of this subject. Based on this principle, the time and spatial development of thermodynamic systems are managed in a manner that the function (Parrondo et al., 2015).

$$\aleph - \quad - \quad , \tag{5.20}$$

Maintains its maximum value in every volume element at all times. As a result, the function remains constant

$$\delta(\sigma - \psi - \phi) = 0. \tag{5.21}$$

This is a fundamental concept in the study of dynamic systems. As a result, for the entire system, one has

$$\delta(\Sigma - \Psi - \Phi) = 0, \tag{5.22}$$

The functions E and V are derived through integration across the entire volume of the system. Equation (5.22) represents the overarching expression of the idea. The entropy produced by Equations (5.20) or (5.21) is determined by the flux potential, which depends on the forces, and the force potential, which depends on the fluxes. All three of them have densities (Andresen et al., 1984).

The combined effect of Equations (5.14), (5.17)–(5.19) along (5.21) represents the application to the governing concept of dissipative processes within linear theory. The equation is derived from the given information.

$$\sum_{i}^{f}\left(\left(J_i - \sum_{j=1}^{i}\frac{L_{ij}+L_{ji}}{2}X_j\right)\delta X_i + \left(X_i - \sum_{j=1}^{f}\frac{R_{ij}+R_{ji}}{2}J_j\right)\delta J_i\right) = 0, L_{ij} \tag{5.23}$$

during which ôXi or ôJi (i = l, 2 f) were independent variations that can be freely selected. As a result, the expressions in parentheses disappear, leading to the reemergence of those linear constitutive equations alongside symmetric coefficient matrices (Varbanov et al., 2020).

5.4. UNIFIED FIELD THEORY OF THERMODYNAMICS

The theory of irreversible procedures within continuous media developed in the 1940s, however, did not possess the same level of consistency and comprehensiveness as the mechanics for continua along with electrodynamics. After extensive research and analysis, the groundbreaking unified theory of fields of thermodynamics was finally published (Ge and Qian, 2016). Within this theory, the comprehensive collection of extensive state factors is considered Ai (i = 1, 2, f) is defined. Their local densities ai moving with velocities represent current densities Ji = api and, of course, are functions of time and position. The balance equations

$$\frac{\partial a_i}{\partial t} = -div\ \ J_i + \sigma_i, \tag{5.24}$$

Consider the values associated with the source densities σi and their corresponding local values. Specifically, let's examine the local entropy density (Klein, 1967).

$$s = s\left(a_1, \ldots, a_i, \ldots, a_f\right) \tag{5.25}$$

Additionally, the function of time and position plays a role. The definition of the intensive variables is as follows:

$$\Gamma_i = \frac{\partial s}{\partial a_i}; \qquad i = 1, 2, \ldots, f. \tag{5.26}$$

The rate for entropy production in a given volume can be determined by taking a time derivative for the entropy and combining that with the balancing equations or the definition for the intensive variables (Haddad, 2017).

$$\sigma = \sum J_i \cdot grad\ \Gamma_i + \sum \Gamma_i \sigma_i, \tag{5.27}$$

The summation is carried out over all values inside the subscript. The term on the right-hand side reflects the production of entropy caused by transport procedures induced by thermodynamic force gradients. The second phrase refers to the entropy creation resulting from the internal generation of substantial quantities, such as the production of certain components or the consumption of others through chemical reactions (Gavassino and Antonelli, 2021).

Tip

Explore the unified field theory of thermodynamics, which seeks to integrate thermodynamic principles with other fields such as statistical mechanics, providing a comprehensive understanding of macroscopic systems.

5.5. LINEARLY INDEPENDENT CHEMICAL REACTIONS AND THEIR ENTROPY PRODUCTION

For a chemical kinetic system consisting of an S elementary reaction

$$\sum_{k=1}^{K} v'_{kt} B_k \to \sum_{k=1}^{K} v''_{kt} B_k \qquad t = 1, 2, ..., S, \tag{5.28}$$

The levels of Ck of component Bk can be altered by the reactions, resulting in their reversal, at a certain pace

$$\sigma_k = \sum_{t=1}^{S} \left(v''_{kt} - v'_{kt} \right) \left(\frac{d\vec{\xi} t}{dt} - \frac{d\overleftarrow{\xi} t}{dt} \right) \qquad k = 1, 2, ..., K, \tag{5.29}$$

The intense parameters include (Rao et al., 2011):

$$\Gamma_k = \frac{\partial s}{\partial_{c_k}} = -\frac{\mu_k}{T}; \qquad k = 1, 2, ..., K. \tag{5.30}$$

Therefore, the overall impact of the entire system for chemical reactions on the generation of entropy is

$$\sigma^{chem} = \sum_{t=1}^{S} \left(\frac{A'_t}{T} - \frac{A''_t}{T} \right) \left(\frac{d\vec{\xi}_t}{dt} - \frac{d\overleftarrow{\xi}_t}{dt} \right) = \sum_{t=1}^{S} \frac{A_t}{T} \frac{d\xi_t}{dt}, \tag{5.31}$$

By mathematical Equations (5.9) and (5.11), A't, A"t, and At (t = 1, 2, S) represent the forward along with backward affinities or affinities associated with the reactions, respectively. Similarly, based on Equations (5.2), (5.6), and (5.7), Œ/dt, d?t/dt, and d4dt (t = 1, 2, S) denote the velocity of the forward and backward reactions, as well as the total reaction rates, accordingly (Tomé and de Oliveira, 2018).

Within the realm of reactions, there exists a group where some reactions are interconnected, while others stand on their own. In the realm of academia, there exists a concept known as stoichiometric independence. This refers to a set of reactions that cannot be derived from any linear combination that includes a subset of the equations within that set. It is a fundamental principle in the field of chemical kinetics and thermodynamics. The relationship between the K x Q matrix, which represents the distinct set of reactions, along the K x S matrix, which represents the overall structure of all reactions, can be described by a set of equations (Ledesma-Durán and Santamaría-Holek, 2022).

$$v'_{kt} = \sum_{u=1}^{Q} \gamma_{tu} v'_{ku}; \quad k = 1, 2, \dots, K; \quad t = 1, 2, \dots, S, \tag{5.32}$$

with

$$\gamma_{tu} = \delta_{tu} + \beta_{tu}, \tag{5.33}$$

where $\beta_{tu}\left(u = 1, 2, \dots, Q; \quad t = 1, 2, \dots, Q, \ Q + 1, \dots, S\right)$ represents the coefficients for a linear combination or

$$\delta_{tu} = \begin{Bmatrix} 1 & u = t \\ 0 & u \neq t \end{Bmatrix} \begin{matrix} t = 1, 2, \dots, Q, \ Q + 1, \dots, S, \\ u = 1, 2, \dots, Q, \end{matrix} \tag{5.34}$$

The Kronecker ô-symbol is a mathematical notation used to represent a discrete function. Similar relationships, similar to equation (5.32), also apply to the elements for [vi'u] and constant coefficients [v].

By replacing the equations that connect the stoichiometric indices for the kinetic system for every reaction with the ones that describe the full set of distinct reactions (Banerjee and Bhattacharyya, 2014).

$$\sum_{t=1}^{S}\left(\frac{A'_t}{T} - \frac{A''_t}{T}\right)\left(\frac{d\vec{\xi}_t}{dt} - \frac{d\overleftarrow{\xi}_t}{dt}\right) = \sum_{u=1}^{Q}\left(\frac{A'_u}{T} - \frac{A''_u}{T}\right)\frac{d\xi^*_u}{dt}, \tag{5.35}$$

With

$$\frac{d\xi^*_u}{dt} = \sum_{t=1}^{S} \gamma_{tu}\left(\frac{d\vec{\xi}_t}{dt} - \frac{d\overleftarrow{\xi}_t}{dt}\right); \qquad u = 1, 2, \dots, Q. \tag{5.36}$$

5.6. EXTENSION BEYOND LOCAL EQUILIBRIUM

It was proposed that while the entire system is in a non-equilibrium state, there exists a state of local cellular equilibrium. Essentially, it was presumed that within all volumes element (cell), the densities to the separate state variables as well as state functions are present assuming the latter are identical functions of the former, just like in equilibrium systems (Rubi et al., 2013).

A significant breakthrough beyond the fundamental idea of local equilibrium was the identification of an additional factor within Fourier's second rule to earn heat conduction. Based on the theory regarding kinetic gases, an equation for motion over temperature is derived, resembling a telegraph-type hyperbolic equation. This equation illustrates the influence of thermodynamic vibrations on the transfer of heat through conduction. According to later research, it was discovered that the additional term in the second Fourier law is related to a force supplement (-grad T) and can be attributed to an entropy term that is proportional to the current density for internal energy flow. This term is included in the equilibrium entropy. Therefore, Cattaneo's theory went beyond the concept of local equilibrium.

After three decades, a new perspective emerged in the field of thermodynamics. Researchers delved into the wave approach, Investigating the entropy density at a local level within an unifying field theory (Beretta et al., 2012). Therefore, the inclusion of terms linear at one of the time derivatives products of the present densities is necessary to expand the thermodynamic factors grad r. Linear relationships are established among the current densities along with the extended forces, using a coefficient matrix that aligns with the united field theory. Ultimately, through the integration of these fundamental equations with local state equations along with balance equations, A set of complex equations like telegraph-type equations has been developed to describe the relationships between intensive factors along with the values of broad variables. These equations explain the principles of thermodynamic waves, specifically the combined effects of wave propagation and conduction. The coefficients in these equations represent various properties such as relaxation times, diffusivities, and conductivities (Bazant, 2013).

5.7. EXPANDING THE SCOPE OF NON-LINEAR PHENOMENOLOGICAL EQUATIONS

In the early days of studying irreversible processes in continua, researchers proposed linear equations to describe the relationship between fluxes and forces. Fourier's law regarding heat conduction and Fick's law for diffusion have been widely accepted and used as reliable approximations for more than a century in the field of academic research (Garcia-Colin and Uribe, 1991). Thus, the assumption of linear constitutive equations seemed to be justified, at least in the case of transport processes. Therefore, it appeared that the assumption for linear constitutive equations was valid, particularly in the context of transport processes. However, a linear function would not accurately represent exponential functions within the Marcelin—Kohnstamm equation. States close to equilibrium will also follow this pattern, as the affinities A' or A" and rates dž/dt along with significant values, while Ä' or dthit = Hence, in such cases,

$$e^{A'/RT} = 1 + \frac{A'}{RT} + \frac{1}{2!}\left(\frac{A'}{RT}\right)^2 + \frac{1}{3}\left(\frac{A'}{RT}\right)^3 + ..., \tag{5.37}$$

The higher-order terms cannot be ignored, and the opposite affinity A' follows the same principle (Gutzow et al., 2008).

It became necessary to generalize the reciprocal relationships for non-linear essential equations due to the limitations of linear constitutive calculations and the requirement of dissipation potentials for the mathematical description of variational principles within thermodynamics (Delsanto et al., 2009). This idea was proposed in various forms

$$\frac{\partial J_i}{\partial X_k} = \frac{\partial J_k}{\partial X_i}; \quad \frac{\partial X_i}{\partial J_k} = \frac{\partial X_k}{\partial J_i}; \quad i,k = 1,2,...,f, \tag{5.38}$$

whereas

$$J_i \equiv J_i\left(X_1,\ldots,X_k,\ldots,X_f\right); \qquad X_i \equiv X_i\left(J_1,\ldots,J_k,\ldots,J_f\right) \tag{5.39}$$

Typically, they exhibit non-linear behavior. Due to the equations, the loss potentials

$$\psi = \psi\left(X_1,\ldots,X_i,\ldots,X_f\right) \tag{5.40}$$

$$\phi = \phi\left(J_1,\ldots,J_i,\ldots,J_f\right) \tag{5.41}$$

There may be situations where this is applicable.

$$J_i = \frac{\partial \psi}{\partial X_i}; \qquad i, = 1, 2, \ldots, f, \tag{5.42}$$

or

$$X_i = \frac{\partial \phi}{\partial J_i}; \qquad i, = 1, 2, \ldots, f, \tag{5.43}$$

Equations (5.38) along with symmetries are closely related, resembling Maxwellian principles (Pekař, 2009).

$$\frac{\partial J_i}{\partial X_k} = \frac{\partial}{\partial X_k}\left(\frac{\partial \psi}{\partial X_i}\right) = \frac{\partial}{\partial X_i}\left(\frac{\partial \psi}{\partial X_k}\right) = \frac{\partial J_k}{\partial X_i} \tag{5.44}$$

$$\frac{\partial X_i}{\partial J_k} = \frac{\partial}{\partial J_k}\left(\frac{\partial \phi}{\partial J_i}\right) = \frac{\partial}{\partial J_i}\left(\frac{\partial \phi}{\partial J_k}\right) = \frac{\partial X_k}{\partial J_i}. \tag{5.45}$$

5.7. DRIVING FORCE FOR CHEMICAL REACTIONS

Over three decades following the proposal of the broader reciprocal relations, numerous efforts were undertaken to demonstrate their validity in the context of a chemical kinetics system involving elementary chemical processes (5.28) and their reversals.

Ultimately, it was demonstrated that the generalized reciprocal relationships will be fulfilled in the chemical kinetic system (5.28) when the appropriate collection is selected (Pross, 2003).

$$\frac{A'_1}{T},\ldots,\frac{A'_u}{T},\ldots,\frac{A'_Q}{T},\ldots,-\frac{A''_1}{T},\ldots,-\frac{A''_u}{T},\ldots,-\frac{A''_Q}{T} \quad (5.46)$$

[refer to the equation on the right-hand side (5.35)] As separate thermodynamic factors [241. Equation (5.36) defines the flux conjugate for both A 'u/T or —A"/T as de/dt. The constitutive equations originate using the Marcelin—Kohnstamm formulas in a specific form (Peka⊠, 2020).

$$\frac{d\xi_u^*}{dt}=\sum_{t=1}^{S}\gamma_{tu}\lambda_t\left[\exp\left(\sum_{v=1}^{Q}\gamma_{tv}A'_v/RT\right)-\exp\left(-\sum_{v=1}^{Q}\gamma_{tv}A''_v/RT\right)\right];\ u=1,2,\ldots,Q. \quad (5.47)$$

Verifying the satisfaction for general reciprocal relationships is straightforward, as the forces within their exponents are now linearly independent. The equations that describe the relationship between different properties

$$\frac{A'_t}{T}=R\ln\left(\frac{d\vec{\xi}_t}{dt}\Big/\lambda_t\right);\ \frac{A''_t}{T}=R\ln\left(\frac{d\overleftarrow{\xi}_t}{dt}\Big/\lambda_t\right);\ t=1,2,\ldots,S. \quad (5.48)$$

We can obtain the desired results by directly inverting equations (5.7) within the form of džt.

Given the distinction between the 2S forces and their independence, it is worth noting that the values of dž/dt; dWdt (t = l, 2, S) remain linearly independent. As a result, the matrix of partial derivatives for the forces about the fluxes takes the form of a diagonal matrix. As a result, the inverse broader relationship will be easily fulfilled. In addition, the principle for dissipative processes has been utilized in the study of systems of elementary chemical reactions in homogenous continua. This has led to the derivation of the kinetic mass action law based on this principle.

The dissipation functions ψ and ϕ can be easily constructed and have the forms

$$\psi = 2R\sum_{t=1}^{S}\lambda_t\left[\exp\left(\sum_{v=1}^{Q}\gamma_{tv}A_v^{'}/RT\right)+\exp\left(-\sum_{v=1}^{Q}\gamma_{tv}A_v^{''}/RT\right)\right] \tag{5.49}$$

thus

$$\phi = 2R\sum_{t=1}^{S}\left[\frac{d\vec{\xi}_t}{dt}\ln\left(\frac{d\vec{\xi}_t}{dt}\Big/\lambda_t\right)+\frac{d\overleftarrow{\xi}_t}{dt}\ln\left(\frac{d\overleftarrow{\xi}_t}{dt}\Big/\lambda_t\right)-\left(\frac{d\vec{\xi}_t}{dt}+\frac{d\overleftarrow{\xi}_t}{dt}\right)\right]. \tag{5.50}$$

The fundamental equation that governs dissipative procedures in the system of basic chemical reactions is derived as follows:

$$\begin{aligned}
&\sum_{u=1}^{Q}\left[\frac{d\xi_u^{*}}{dt}-2\sum_{t=1}^{S}\gamma_{tu}\lambda_t\exp\left(\sum_{v=1}^{Q}\gamma_{tv}A_v^{'}/RT\right)\right]\delta\left(\frac{A'_u}{T}\right)\\
&+\sum_{u=1}^{Q}\left[\frac{d\xi_u^{*}}{dt}+2\sum_{t=1}^{S}\gamma_{tu}\lambda_t\exp\left(-\sum_{v=1}^{Q}\gamma_{tv}A_v^{''}/RT\right)\right]\delta\left(\frac{A''_u}{T}\right)\\
&+\sum_{t=1}^{S}\left[\frac{A'_t}{T}-\frac{A''_t}{T}-2R\ln\left(\frac{d\vec{\xi}_t}{dt}\Big/\lambda_t\right)\right]\delta\left(\frac{d\vec{\xi}_t}{dt}\right)\\
&+\sum_{t=1}^{S}\left[-\frac{A'_t}{T}+\frac{A''_t}{T}-2R\ln\left(\frac{d\overleftarrow{\xi}_t}{dt}\Big/\lambda_t\right)\right]\delta\left(\frac{d\overleftarrow{\xi}_t}{dt}\right)=0
\end{aligned} \tag{5.51}$$

The variation was conducted about the whole set of separate variables. Thus, based on Equation (5.51), all expressions within the brackets must be zero. By combining the obtained equations, one can derive the constitutive calculations (5.47) and (5.48). In ideal systems, these equations are arcane equivalent to the Guldberg-Waage kinetic mass motion law, as demonstrated here. This allows us to deduce the basic formula of chemical kinetics from the regulating theory of dissipative procedures. Furthermore, a study revealed that in the case of non-ideal systems, the deduction mentioned earlier can be extended into the kinetics mass action law within a more generalized form (Schiffer et al., 2021).

$$\frac{d\xi_t}{dt}=\frac{d\vec{\xi}_t}{dt}-\frac{d\overleftarrow{\xi}_t}{dt}=\overleftarrow{k}_t\prod_{k=1}^{k}a_k^{v_{kt}^{'}}-\overleftarrow{k}_t\prod_{k=1}^{K}a^{v_{kt}^{''}};\qquad t=1,2,\ldots,S, \tag{5.52}$$

whereas

$$a_k = \exp\left(\frac{\mu_k - \mu_k^\theta}{RT}\right); \quad k = 1, 2, \ldots, K, \tag{5.53}$$

The relative operations of each element in the system should not be confused with the values in equations (5.25) or (5.55), and so forth In the comprehensive field theory of thermodynamics, the definitions of chemical potentials along with the density associated with the Gibbs function are established.

$$\mu_k \equiv \left(\frac{\partial g}{\partial c_k}\right) T, p, c_{k \neq j} \tag{5.54}$$

$$g \equiv g(r,t) = g\left[a_1(r,\ t), \ldots, a_i(r,\ t), \ldots, a_f(r,\ t)\right]. \tag{5.55}$$

What is the process by which the chemical kinetic system evolves?

When studying chemical kinetics, it is possible to express the concentrations for the components within the kinetic mass-action law by considering the extent of the reaction, as long as the system remains closed. Therefore, we can deduce a differential equation with just one aspect: the extent to which the reaction. The solution for this equation represents the progression of the reaction over time.

Another possibility for studying the evolution is offered by thermodynamics. Through the combination of various equations, we can derive differential equations that describe the evolution of chemical reactions over time. These equations involve thermodynamic forces and reaction rates, providing valuable insights into the reaction process (Goto et al., 1990). When we differentiate the De Donder relationship concerning time, one obtains

$$\frac{d(A/T)}{dt} = s''\frac{d\xi}{dt}, \tag{5.56}$$

in which the coefficient

$$s'' = \frac{\partial^2 s}{\partial \xi^2}, \tag{5.57}$$

Typically, the outcome is influenced by the progression of the reaction along with the duration it takes. In the context of linear approximation, the value of s" remains constant. Equation (5.56) may be divided into two parts, one for the forward reaction and another for the reverse reaction:

$$\frac{d(A'/T)}{dt} = s''\frac{d\vec{\xi}}{dt} \tag{5.56'}$$

or

$$\frac{d(A''/T)}{dt} = s''\frac{d\bar{\xi}}{dt}. \tag{5.56''}$$

These equations may be combined to further analyze the system (Politzer et al., 2014).

$$\frac{d\bar{\xi}}{dt} = \lambda \exp\left(\frac{A'}{T}\Big/R\right) \tag{5.58'}$$

and

$$\frac{d\bar{\xi}}{dt} = \lambda \exp\left(\frac{A''}{T}\Big/R\right) \tag{5.58''}$$

one obtains

$$\frac{d(A'/T)}{dt} = s''\lambda \exp\left(\frac{A'}{T}\Big/R\right) \tag{5.59'}$$

and

$$\frac{d(A''/T)}{dt} = s''\lambda \exp\left(\frac{A''}{T}\Big/R\right) \tag{5.59''}$$

or

$$\frac{1}{\bar{r}^2}\frac{d\bar{r}}{dt} = \frac{s''}{R} \tag{5.60'}$$

and

$$\frac{1}{\bar{r}^2}\frac{d\bar{r}}{dt} = \frac{s''}{R} \tag{5.60''}$$

As one delves into the realm of reaction rates

$$\vec{r} = \frac{d\vec{\xi}}{dt} \qquad \textit{and} \qquad \bar{r}\frac{d\bar{\xi}}{dt} \tag{5.61}$$

and the forces A'/T along with A"/T.

By making the assumption of constancy for s," one can derive the solutions using linear approximation.

$$\frac{A'}{T} = \left(\frac{A'}{T}\right)_0 - R\ln\left[1 - t\frac{s''\lambda}{R}\exp\left[\left(\frac{A'}{T}\right)_0\Big/R\right]\right] \tag{5.62'}$$

$$\frac{A''}{T} = \left(\frac{A''}{T}\right)_0 - R\ln\left[1 - t\frac{s''\lambda}{R}\exp\left[\left(\frac{A''}{T}\right)_0 \Big/ R\right]\right], \tag{5.62"}$$

The set of equations (5.59) along with their corresponding solutions

$$\vec{r} = \vec{r}_0 \big/ \left(1 - ts''\vec{r}_0 / R\right) \tag{5.63'}$$

$$\bar{r} = \bar{r}_0 \big/ \left(1 - ts''\bar{r}_0 / R\right) \tag{5.63"}$$

The system consists of 60 equations that explain the reversible basic chemical process over time.

$$\sum_{k=1}^{K} v_k B_k \rightleftarrows \sum_{k=1}^{K} v_k B_k \tag{5.64}$$

Among the different components, B, B, and BK are involved. Therefore, it is evident the way the affinities A' or A" and reaction rates depend on time. The indices within equations (5.62) or (5.63) denote the initial values.

Through the integration of the disparity between equations (5.63') and (5.63"), one can take our understanding one step further. One acquires

$$\xi = \int_0^t \left(\frac{d\vec{\xi}}{dt} - \frac{d\bar{\xi}}{dt}\right) dt = \ln\frac{1-\left(s''\vec{r}_0/R\right)t}{1-\left(s''\bar{r}_0/R\right)t}, \tag{5.65}$$

What is the solution to the mass action law for a reversible elementary reaction? By comparing it to the typical solution within standard chemical kinetics, one can understand the importance of the constants." As an illustration, consider the scenario of a reversible elementary reaction (Peng et al., 2020).

$$B_1 + B_2 \rightleftharpoons B_3 + B_4 \tag{5.66}$$

The De Donder relations can be combined alongside the Marcelin—Kohnstamm equations to analyze elementary reversible reactions within chemical kinetics systems,

For instance, to the system expressed in equations (5.28). Typically, though,

$$s'' = \frac{\partial}{\partial\xi}\left(\frac{\partial s}{\partial\xi}\right) = \frac{\partial\left(A/T\right)}{\partial\xi}$$

The value that is given is changeable and dependent on the factors A/T and an additional component. Hence, to accomplish the integration with equations (5.59) along (5.60), one must possess knowledge of these functions (Beard and Qian, 2007).

5.8. THE SIGNIFICANCE OF SYMMETRY WITHIN THE ANALYSIS OF IRREVERSIBLE PHENOMENA

Symmetry plays a crucial role in fundamental equations, also known as experiential equations. The matrix of coefficients

$$\begin{bmatrix} L_{11} & L_{12} & L_{13} & \cdots & L_{1f} \\ L_{21} & L_{22} & L_{23} & \cdots & L_{2f} \\ L_{31} & L_{32} & L_{33} & \cdots & L_{3f} \\ . & . & . & & . \\ . & . & . & & . \\ . & . & . & & . \\ L_{f1} & L_{f2} & L_{f3} & \cdots & L_{ff} \end{bmatrix} \tag{6.67}$$

It remains symmetrical under all circumstances (Luzzi et al., 2013). In the linear theory, elements such as Lik are fixed coefficients associated with linear constitutive equations. In the broader context of the non-linear theory, every component of the matrix mentioned above is described as

$$L_{ij} = \frac{\partial J_i}{\partial X_k}; \qquad i,k = 1,2,...,f. \tag{5.68}$$

The presence of symmetry in both cases was crucial in establishing the presence of dissipation functions. This, in turn, allowed for the derivation of the corresponding constitutive equations through the fundamental variational law of thermodynamics (Astumian, 2018). Constitutive equations are derived from various fields of study, initially unrelated to thermodynamics. Here are some examples of essential equations in various scientific fields: Hooke's law (1660) for elasticity, the law of Ohm (1827) during electric transmission, Fourier's law (1822) over conduction of heat, the Navier—Stokes formulas (1845) in viscous liquids, Fick's law (1855) in diffusion, along with a Guldberg—Waage kinetic motion law (1867) over chemical kinetics (Eu, 1984).

The derivation for the essential equations of various phenomena, such as diffusion, heat conduction, flow, and turbulence, using the variational principle known as thermodynamics leads to a synthesis of material along with heat transfer, rheology, and thermodynamics (Polettini and Esposito, 2014). The study of chemical reactions has historically been divided into two branches: chemical thermodynamics and chemical kinetics (Eu, 1980). The symmetrical nature found within generalized reciprocal relations allows for the integration of two scientific disciplines, where chemical reactions can be studied using the principles of process thermodynamics along chemical kinetics, respectively. Therefore, symmetry can be seen as a tool for combining these two fields of study. It could be argued that symmetry possesses a universal characteristic of serving as a tool for synthesis and integration (Razdan et al., 2023).

ACTIVITY 5.1.

Conduct experiments to identify and quantify the driving forces behind chemical reactions. Investigate factors such as temperature, concentration, and catalysts to observe their effects on reaction rates.

SUMMARY

- Chapter Five, "Chemical Kinetics and Thermodynamics," delves into the fundamental principles governing chemical reactions and thermodynamic processes. It begins by elucidating the concept of chemical affinity, which underpins both modern chemical kinetics and thermodynamics. The chapter progresses to outline the procedural aspects of chemical reactions and the intricate processes of thermodynamics, including a comprehensive discussion on the unified field theory that integrates these phenomena.
- Furthermore, the chapter explores the concept of linearly independent chemical reactions and their role in entropy production. It extends beyond local equilibrium to discuss the driving forces behind chemical reactions, emphasizing the significance of symmetry in analyzing irreversible phenomena.

REVIEW QUESTIONS

1. What is chemical affinity, and how does it relate to both chemical kinetics and thermodynamics?
2. Explain the unified field theory of thermodynamics and its significance in understanding chemical processes.
3. Discuss the procedure of a chemical reaction and the factors that influence reaction rates.
4. How does thermodynamics describe energy changes during chemical reactions? Provide examples.
5. What are linearly independent chemical reactions, and why is entropy production important in their analysis?
6. Compare and contrast local equilibrium and extensions beyond local equilibrium in chemical systems.
7. Why is symmetry analysis crucial in the study of irreversible phenomena in chemical kinetics?

MULTIPLE CHOICE QUESTIONS

1. Which concept links modern chemical kinetics and thermodynamics?

a. Chemical equilibrium

b. Chemical affinity

c. Reaction rate

d. Entropy

2. What does the unified field theory of thermodynamics aim to unify?

a. Quantum mechanics and thermodynamics

b. Chemical kinetics and thermodynamics

c. Classical mechanics and thermodynamics
d. Special relativity and thermodynamics

3. The driving force for chemical reactions is primarily determined by:

a. Temperature change
b. Entropy change
c. Pressure change
d. Enthalpy change

4. Which term describes chemical reactions that occur without external intervention?

a. Homogeneous reactions
b. Irreversible reactions
c. Autocatalytic reactions
d. Spontaneous reactions

5. In chemical kinetics, what does the term "activation energy" refer to?

a. Energy released during a reaction
b. Energy barrier for the reaction to occur
c. Heat absorbed during a reaction
d. Energy difference between products and reactants

6. Which principle governs the equilibrium constant of a chemical reaction?

a. Le Chatelier's principle
b. The Arrhenius equation
c. The Law of Mass Action
d. Hess's law

7. Why is the concept of symmetry crucial in analyzing irreversible processes in thermodynamics?

a. It determines the spontaneity of reactions.
b. It explains the direction of energy flow.
c. It quantifies entropy changes.
d. It predicts reaction rates.

Answers to Multiple Questions

1. (b) 2. (b) 3. (d) 4. (c) 5. (b)
6. (c) 7. (b)

REFERENCES

1. Alonso, A. A., & Ydstie, B. E. (1996). Process systems, passivity and the second law of thermodynamics. *Computers & Chemical Engineering, 20*, 1119–1124.
2. Andresen, B., Berry, R. S., Ondrechen, M. J., & Salamon, P. (1984). Thermodynamics for processes in finite time. *Accounts of Chemical Research, 17*(8), 266–271.
3. Aquilanti, V., Borges, E. P., Coutinho, N. D., Mundim, K. C., & Carvalho-Silva, V. H. (2018). From statistical thermodynamics to molecular kinetics: The change, the chance and the choice. *Rendiconti Lincei. Scienze Fisiche e Naturali, 29*, 787–802.
4. Astumian, R. D. (2018). Trajectory and cycle-based thermodynamics and kinetics of molecular machines: The importance of microscopic reversibility. *Accounts of Chemical Research, 51*(11), 2653–2661.
5. Banerjee, K., & Bhattacharyya, K. (2014). Entropy production as a universal functional of reaction rate: Chemical networks close to steady states. *Journal of Mathematical Chemistry, 52*, 820–832.
6. Bazant, M. Z. (2013). Theory of chemical kinetics and charge transfer based on nonequilibrium thermodynamics. *Accounts of Chemical Research, 46*(5), 1144–1160.
7. Beard, D. A., & Qian, H. (2007). Relationship between thermodynamic driving force and one-way fluxes in reversible processes. *PloS One, 2*(1), 144.
8. Bejan, A. (2017). Evolution in thermodynamics. *Applied Physics Reviews, 4*(1), 254–256.
9. Beretta, G. P., Keck, J. C., Janbozorgi, M., & Metghalchi, H. (2012). The rate-controlled constrained-equilibrium approach to far-from-local-equilibrium thermodynamics. *Entropy, 14*(2), 92–130.
10. Berg, C. A. (1980). Process integration and the second law of thermodynamics: Future possibilities. *Energy, 5*(8, 9), 733–742.
11. Chen, C. C., & Mathias, P. M. (2002). Applied thermodynamics for process modeling. *AIChE Journal, 48*(2), 194.
12. Christine King, M. (1981). Experiments with time: Progress and problems in the development of chemical kinetics. *Ambix, 28*(2), 70–82.
13. Cioc, R. C., Crockatt, M., van der Waal, J. C., & Bruijnincx, P. C. (2022). The interplay between kinetics and thermodynamics in furan Diels–Alder chemistry for sustainable chemicals production. *Angewandte Chemie International Edition, 61*(17), 202114720.
14. Delsanto, P. P., Gliozzi, A. S., Bruno, C. L., Pugno, N., & Carpinteri, A. (2009). Scaling laws and fractality in the framework of a phenomenological approach. *Chaos, Solitons & Fractals, 41*(5), 2782–2786.
15. Eu, B. C. (1980). A modified moment method and irreversible thermodynamics. *The Journal of Chemical Physics, 73*(6), 2958–2969.
16. Eu, B. C. (1984). Kinetic theory and irreversible thermodynamics of nonlinear transport processes in quantum systems. *The Journal of Chemical Physics, 80*(5), 2123–2140.

17. Eu, B. C. (2002). *Generalized Thermodynamics: Thermodynamics of Irreversible Processes and Generalized Hydrodynamics.* Springer Science & Business Media.

18. Garcia-Colin, L. S., & Uribe, F. J. (1991). Extended irreversible thermodynamics beyond the linear regime: A critical overview. *Journal of Non-Equilibrium Thermodynamics, 16*(2), 89–128.

19. Garfinkle, M. (2000). The thermodynamic natural path in chemical reaction kinetics. *Discrete Dynamics in Nature and Society, 4*(2), 145–164.

20. Gavassino, L., & Antonelli, M. (2021). Unified extended irreversible thermodynamics and the stability of relativistic theories for dissipation. *Frontiers in Astronomy and Space Sciences, 8*, 686344.

21. Ge, H., & Qian, H. (2016). Mesoscopic kinetic basis of macroscopic chemical thermodynamics: A mathematical theory. *Physical Review E, 94*(5), 052150.

22. Goto, M., Smith, J. M., & McCoy, B. J. (1990). Parabolic profile approximation (linear driving-force model) for chemical reactions. *Chemical Engineering Science, 45*(2), 443–448.

23. Gutzow, I., Schmelzer, J. W. P., & Petroff, B. (2008). Phenomenological theories of glass transition: Classical approaches, new solutions and perspectives. *Journal of Non-Crystalline Solids, 354*(2–9), 311–324.

24. Haddad, W. M. (2017). Thermodynamics: The unique universal science. *Entropy, 19*(11), 621.

25. Hutter, K. (1977). The foundations of thermodynamics, its basic postulates and implications: A review of modern thermodynamics. *Acta Mechanica, 27*(1), 1–54.

26. Ishida, M., & Chuang, C. C. (1997). New approach to thermodynamics. *Energy Conversion and Management, 38*(15–17), 1543–1555.

27. Kaganovich, B. M., Keiko, A. V., Shamansky, V. A., & Zarodnyuk, M. S. (2012). On the interrelations between kinetics and thermodynamics as the theories of trajectories and states. Intech, Rijeka.

28. Keizer, J. (1982). Nonequilibrium statistical thermodynamics and the effect of diffusion on chemical reaction rates. *The Journal of Physical Chemistry, 86*(26), 5052–5067.

29. Klein, M. J. (1967). Thermodynamics in Einstein's thought: Thermodynamics played a special role in Einstein's early search for a unified foundation of physics. *Science, 157*(3788), 509–516.

30. Kondepudi, D. K. (2008). *Introduction to Modern Thermodynamics.* Chichester: Wiley.

31. Ledesma-Durán, A., & Santamaría-Holek, I. (2022). Energy and entropy in open and irreversible chemical reaction–diffusion systems with asymptotic stability. *Journal of Non-Equilibrium Thermodynamics, 47*(3), 311–328.

32. Luzzi, R., Vasconcellos, A. R., & Ramos, J. G. (2013). *Statistical Foundations of Irreversible Thermodynamics.* Springer Science & Business Media.

33. Parrondo, J. M., Horowitz, J. M., & Sagawa, T. (2015). Thermodynamics of information. *Nature Physics, 11*(2), 131–139.

34. Pekař, M. (2009). Thermodynamic framework for design of reaction rate equations and schemes. *Collection of Czechoslovak Chemical Communications, 74*(9), 1375–1401.
35. Pekař, M. (2020). Thermodynamic driving forces and chemical reaction fluxes; Reflections on the steady state. *Molecules, 25*(3), 699.
36. Peng, Y., Qian, H., Beard, D. A., & Ge, H. (2020). Universal relation between thermodynamic driving force and one-way fluxes in a nonequilibrium chemical reaction with complex mechanism. *Physical Review Research, 2*(3), 033089.
37. Polettini, M., & Esposito, M. (2014). Irreversible thermodynamics of open chemical networks. I. Emergent cycles and broken conservation laws. *The Journal of Chemical Physics, 141*(2), 658–664.
38. Politzer, P., Murray, J. S., Yepes, D., & Jaque, P. (2014). Driving and retarding forces in a chemical reaction. *Journal of Molecular Modeling, 20*, 1–6.
39. Pross, A. (2003). The driving force for life's emergence: Kinetic and thermodynamic considerations. *Journal of Theoretical Biology, 220*(3), 393–406.
40. Rao, T., Xiao, T., & Hou, Z. (2011). Entropy production in a mesoscopic chemical reaction system with oscillatory and excitable dynamics. *The Journal of Chemical Physics, 134*(21), 65–70.
41. Razdan, N. K., Lin, T. C., & Bhan, A. (2023). Concepts relevant for the kinetic analysis of reversible reaction systems. *Chemical Reviews, 123*(6), 2950–3006.
42. Rubi, J. M., Bedeaux, D., Kjelstrup, S., & Pagonabarraga, I. (2013). Chemical cycle kinetics: Removing the limitation of linearity of a non-equilibrium thermodynamic description. *International Journal of Thermophysics, 34*, 1214–1228.
43. Santillán, M. (2014). *Chemical Kinetics, Stochastic Processes, and Irreversible Thermodynamics.* Heidelberg: Springer.
44. Schiffer, Z. J., Limaye, A. M., & Manthiram, K. (2021). Thermodynamic discrimination between energy sources for chemical reactions. *Joule, 5*(1), 135–148.
45. Tomé, T., & de Oliveira, M. J. (2018). Stochastic thermodynamics and entropy production of chemical reaction systems. *The Journal of Chemical Physics, 148*(22), 325–330.
46. Varbanov, P. S., Chin, H. H., Plesu Popescu, A. E., & Boldyryev, S. (2020). Thermodynamics-based process sustainability evaluation. *Energies, 13*(9), 2132.
47. Wang, J. (2009). Modern thermodynamics–New concepts based on the second law of thermodynamics. *Progress in Natural Science, 19*(1), 125–135.
48. White, W. B. (1997). Thermodynamic equilibrium, kinetics, activation barriers, and reaction mechanisms for chemical reactions in karst terrains. *Environmental Geology, 30*, 46–58.
49. Yu, Y., & Zhang, Y. X. (2017). Coupling of chemical kinetics and thermodynamics for simulations of leaching of cement paste in ammonium nitrate solution. *Cement and Concrete Research, 95*, 95–107.

CHAPTER 6

Electrochemistry

LEARNING OBJECTIVES

After studying this chapter, you will be able to:

1 Understand the fundamental terminologies used in electrochemistry.

2 Explain the process and principles of electrolysis.

3 Apply Faraday's laws of electrolysis to calculate quantitative aspects of electrochemical reactions.

4 Evaluate the Arrhenius theory of electrolytic dissociation and its limitations.

5 Describe the different types of electrical conductance and their measurements.

6 Discuss Kohlrausch's law and its significance in determining molar conductivities.

7 Analyze the principles and components of electrochemical cells, including electrolytic cells and galvanic cells.

8 Explain the phenomena of corrosion and methods to prevent it in different environments.

KEY TERMS FROM THIS CHAPTER

Arrhenius theory
Electrical conductance
Electrolysis
Kohlrausch's law
Corrosion
Electrochemical cells
Faraday's laws of electrolysis

INTRODUCTION

Electrochemistry is a fascinating field of study that focuses on the conversion of electrical energy into chemical energies into. One will cover reaction of electrolyte, which are reactions which take place when electricity travels through solutions, and electromagnetic reactions, which are reactions which generate electric energy. There are various instances of electrolytic reactions, such as electrolysis, electroplating, and electrorefining of metals. There are various examples of electrochemical reactions, such as those that take place in cells or batteries (Wagner et al., 2010).

6.1. TERMINOLOGIES IN ELECTROCHEMISTRY

Here are some key terms commonly used in the field of electrochemistry:

Electricity conductors are materials that promote the transport of electrons. Conductive materials can be broadly classified into two main groups: electronic conductors and electrolytic conductors.

Table 6.1. *Tabular Representation of Understanding the Distinction in Electrolytic Conductors and Electronic Conductors*

	Metallic or Electronic Conductors	Solution or Electrolytic Conductors
1.	Electric current flows by the passage of electrons across the metallic lattice, including copper (Cu) or silver (Ag).	Ionic current can be conducted in a state of melting or through aqueous solutions with electrolytes, like NaCl (aq) and NaCl (fused).
2.	The passage pertains to the present outcomes that are exclusively related to bodily changes.	The flow of electric current causes both chemical and physical changes.
3.	Typically, there is none exchange of substances observed.	It encompasses the movement on matter within the shape of ions.
4.	Typically, resistance tends to rise as current passes through, which can be attributed to a rise in temperature. The flow from electrons is hindered by the increased thermal movement of metal ions just like the temperature rises.	Typically, resistance tends to decrease when the temperature increases. This is because the medium becomes less viscous, and the ions become less hydrated.
5.	Metals generally have a high level of conductivity.	The conductivity of electrolysis has a comparatively low conducting capacity.

(*Source:* Schröder et al, Creative Commons License).

1. **Insulators:** It comprise materials that impede the movement of electrons. Examples encompass materials such as silk, wool, wood, plastic, and others.
2. **Electrolytes:** An electrolyte seems to be a substance which, when in solution or within the molten state, can conduct electricity and undergo simultaneous decomposition. The extent and degree of separation of various electrolytes in solution varies. There are two broad categories of electrolytes: Understanding the differences between weak and strong electrolytes.
3. **Strong Electrolytes:** Substances that easily dissociate and create a highly conductive liquid when mixed with water are known as strong electrolytes. Examples of strong electrolytes include salts (except), acidic minerals such as HCl, tc, and bases such as NaOH, KOH, and others. Strong electrolytes are nearly completely ionized when diluted at normal levels (Garnett et al., 1990).
4. **Weak Electrolytes:** Some substances only partially dissociate in water, resulting

Remember

Remember that electrochemistry deals with the study of chemical processes that involve the movement of electrons.

in low conductivity. These substances are known as weak electrolytes. Examples include inorganic acids such as , in addition to bases such as and amines.

5. **Electrodes:** To allow the flow of current across an electrolytic conductor, it have two rods and plates connected to the terminals for a battery. These rods/plates are known as electrodes. The electrode responsible for the process of oxidation is referred to as the anode, while the electrode responsible for the reduction reaction is known as the cathode (Thévenot et al., 2001).

6.2. ELECTROLYSIS

Electrolysis (Figure 6.1) is a process that occurs by passing of electric current through an electrolyte, whether it is in a molten or just solution form. This results in intriguing chemical changes, including the reduction for ions within electronation and the oxidation via de-electronation.

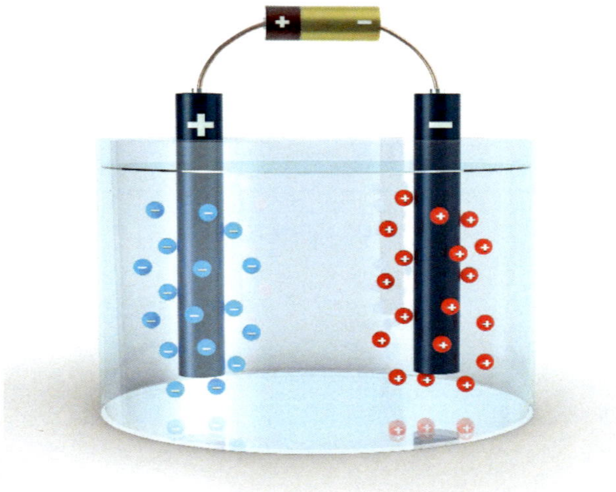

Figure 6.1. *Image of electrolysis.*

(Source: online learning, Creative Commons License).

When an electrolytic solution contains multiple ions and undergoes electrolysis, it is noticed that none of the ion's discharge via the electrodes at the same time. Theory of preferential discharge explains the release of ions through electrodes (Table 6.2). According to the information provided, when multiple types of ions are drawn to an electrode, the ion being discharged is the one that needs the least amount of energy. The voltage that occurs when an ion is discharged and deposited onto the correct electrode is referred to as a discharge potentially or deposit potential. Ionic discharge potentials exhibit variability (Berzins and Delahay, 1953).

Table 6.2. *Tabular Data of Instances of Preferred Discharge Theory*

Electrolyte	Electrode	Cathodic Reaction	Anodic Reaction
Aqueous acidified $CuCl_2$ solution	Pt	$Cu^{2+} + 2e^- \rightarrow Cu$	$2Cl^- \rightarrow +Cl_2 \quad 2e^-$
Molten $PbBr_2$	Pt	$Pb^{2+} + 2e^- \rightarrow Pb$	$2Br^- \rightarrow +Br_2 \quad 2e^-$
Sodium chloride solution	Hg	$2Na^+ + 2e^- \rightarrow 2Na$	$2Cl^- \rightarrow +Cl_2 \quad 2e^-$
Silver nitrate solution	Pt	$Ag^+ + e^- \rightarrow Ag$	$2OH^- \rightarrow \frac{1}{2}O_2 + H\ O_2 + 2e^-$
Sodium nitrate solution	Pt	$2H^+ + 2e^- \rightarrow H_2$	$2OH^- \rightarrow \frac{1}{2}O_2 + H\ O_2 + 2e^-$

(*Source:* Luman Learning, Creative Commons License).

6.3. FARADAY'S LAW OF ELECTROLYSIS

Faraday introduced the 'laws of electrolysis' in 1834, which describe the correlation within the amount of electric charge transferred by an electrolyte with the quantity of material deposited across the electrodes (Mazloomi et al., 2013).

6.3.1. Faraday's First Law

The amount of material deposited in the electrolyte is exactly proportional with the electric charge passing through it when electrical current is applied.

W represents the substance mass placed via enacting coulomb Q for charge, thus law states that there is a relation between the two (Palit, 1967):

$$W \propto Q$$

$$Q = I \times t$$

$$W = Z \times Q$$

Z denotes a constant, referred to as the electrical equivalent, it is specific to the substance getting deposited. When an electric current for one ampere passes through a circuit for a duration of a second, resulting in a charge of one coulomb (Q = 1), the outcome is that the work done (W) is equal to the impedance (Z).

The electrical equivalence refers to the amount of mass in a certain substance that is deposited when a certain amount of charge or current is applied for a specific duration of time (Smolinka et al., 2022).

6.3.2. Faraday's Second Law

When an equal amount of charge passes through various electrolytes, the masses of the various substances deposited on the corresponding electrodes are proportional to their similar masses. Once again, in accordance with the first law,

$$W = Z \times Q$$

When the charge, Q, is equal to 96500 coulombs, the work, W, is equivalent to the gram equivalent mass, E (Daniell, 1840).

Therefore,

$$E = Z \times 96500$$

$$Z = \frac{E}{96500}$$

$$Z1 = \frac{E1}{E2}$$

6.4. ARRHENIUS THEORY OF ELECTROLYTIC DISSOCIATION

In Arrhenius developed an extensive framework to explain the characteristics of electrolytic solutions. The theory's primary postulates are:

a. When dissolving in water, an electrolyte separates into two kinds of particles with charges, one with a positive charge or the other with a negative charge. These particles with electric charge are known as ions. Ions with a positive charge are known as cations, while ions with a negative charge are referred to as anions (Arrhenius, 1889).

$$A^{+}B^{-} + aq. \longrightarrow A^{+}(aq.) + B^{-}(aq.)$$

b. Ionization seems the process of separating molecules through ions of an electrolyte. The degree of ionization process or degree of separation refers to the portion of the total amount of molecules within a solution that exist as ions. It is represented by the symbol 'α.'

$$\alpha = \frac{\textbf{Number of molecules dissociated into ions}}{\textbf{Toal number of molecules}}$$

c. It is a constant process of ions in solution coming together to create neutral molecules, resulting in a dynamic equilibrium within ionized or non-ionized molecules.

$$AB \rightleftharpoons A^{+} + B^{-}$$

d. Applying the principle for mass action to a situation over equilibrium

$$\frac{[A^{+}][B^{-}]}{[AB]} = K$$

K is commonly referred to as the ionization constant. Electrolytes with a high value for K are referred to as strong electrolytes, while those with low levels of K are known to be weak electrolytes.

a. While a voltage is applied to the electrolyte solution, the positively charged ions (cations) are attracted to the cathode while the ions that are negative (anions) are attracted to the anode. As a result, the ions undergo discharge, leading to the process of electrolysis. The ions always get discharged within equivalent amounts, regardless of their respective speeds (Bjerrum, 1909).

b. The solution of electrolysis is typically neutral because the total charge upon a particular set of ions remains balanced by the overall charge of opposite ions set. But not always needed for the ions number in two equal sets.

Did you know?

Did you know that while the Arrhenius theory explains electrolytic dissociation for many substances, it doesn't account for substances that dissociate without water?

$$AB \rightleftharpoons A^{+} + B^{-} \qquad (Both\ ions\ are\ equal)$$

$$NaCl \rightleftharpoons Na^{+} + Cl^{-} \qquad (Both\ ions\ are\ equal)$$

$$AB_2 \rightleftharpoons A^{2+} + 2B^{-} \qquad (Anions\ are\ double\ that\ of\ cations)$$

$$BaCl_2 \rightleftharpoons Ba^{2+} + 2Cl^{-} \qquad (Anions\ are\ double\ that\ of\ cations)$$

$$A_2B \rightleftharpoons 2a^{+} + B2^{-} \qquad (Cations\ are\ double\ that\ of\ anions)$$

$$Na_2SO_4 \rightleftharpoons 2Na^{+} + SO_4^{2-} \qquad (Cations\ are\ double\ that\ of\ anions)$$

The characteristics of electrolytes within a solution are determined by the ions found in the solution. As an illustration, an acidic solution consistently contains H+ ions, whereas a solution that is basic has OH– ions. The distinctive features for solutions are attributed to the presence of H+ ions as well OH– ions, accordingly (Masa et al., 2019).

6.4.1. Limitations for Arrhenius Theory

i. Applying Ostwald's diluting law, that relies on Arrhenius theory, is not feasible for strong electrolytes.

ii. Electric current can be conducted by strong electrolytes even without the presence of water, such as in a fused state. This statement contradicts the Arrhenius principle, which emphasizes the necessity of a solvent over ionization to occur.

iii. The Arrhenius theory presumes that ions exist independently, but it does not consider the factors that affect the movement of the ions (Gewald et al., 2019).

6.4.2. Factors Affecting the Extent of Ionization

i. Solutes Nature: Ionizable components of an element molecule are predominantly bound through covalent bonds instead of electrovalent bonds, the number of ions provided in solution is reduced.

ii. Nature of solvent: The primary role of a solvent serves to diminish electrostatic charges of attraction within the two ions to establish a separation between them.

iii. Dilution: The concentration for a solution has an inverse relationship with the level of ionization (Figure 6.2) caused by an electrolyte. Therefore, the level of ionization rises as the solution becomes more diluted, meaning the solution concentration decreases.

iv. Temperature: As temperature rises, the extent for ionization also increases. This phenomenon occurs because of increased temperatures, leading to increased molecular speed which surpasses the forces for attraction among ions (Jockusch et al., 1997).

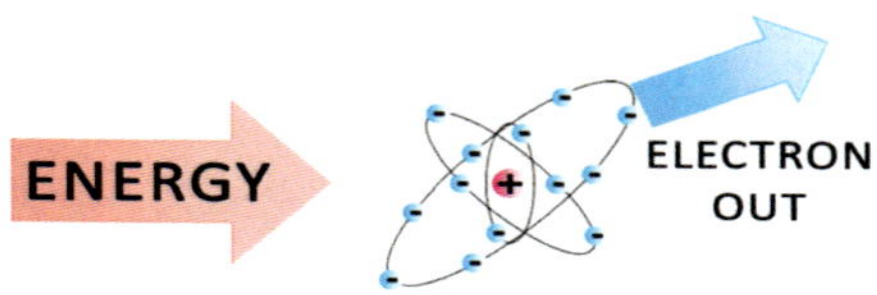

Figure 6.2. *Image of ionization.*

(Source: Energy Education, Creative Commons License).

6.5. ELECTRICAL CONDUCTANCE

Conductivity is a crucial characteristic of conductors, both metals and electrolytic, as it allows for the smooth flow of electricity. This is the inverse of resistance, meaning

$$\text{Conductance} = G = \frac{1}{Resistance} = \frac{1}{R} \qquad \text{(i)}$$

It is measured in a unit known as the reciprocal Siemens or ohms (ohm–1 or mho) (Laity, 1959).

6.5.1. Conductivity or Specific Conductance

Resistance of a conductor changes in proportion to its length (l) or contrary to its cross-sectional area.

$$i.e.\ R \propto \frac{I}{a} \quad or \quad R = \rho\frac{I}{a} \qquad \text{(ii)}$$

Where, ρ is referred to as the specific resistance. If a 1cm length and of $1cm^2$ area the resistance ® is equal to the resistivity ρ

The resistance is regarded as a resistance of one cubic centimeter for a conductor.

The conductance seems to be a reciprocal of a particular resistance, representing the electrical conductivity of one cubic centimeter of a conductor. Shown by κ,

—

(iii)

where,   kappa is the conductance with specified characteristics.

Conductance × Cell constant = Specific conductance.   (iv)

Conductivity is synonymous with conductance (Cox et al., 2007).

By eq. (ii), one has

$$\rho = \frac{a}{I} \cdot R \qquad or \qquad \frac{1}{\rho} = \frac{I}{a} \cdot \frac{1}{R}; \qquad \kappa = \frac{I}{a} \times C \left(\frac{I}{a} = cell\ constant \right)$$

6.5.2. Equivalent Conductance

Equivalent conductivity refers to the level of conductance shown by each individual ion generated from the gram-equivalent of an electrolyte within a certain solution.

The symbol Λ represents it.

Generally, Λ=κ× V (v)

Assuming a volume for V in mL that contains 1 g-equivalent of an electrolytes

If the amount of a solution is expressed as c g-equivalents per liter, the quantity needed to hold One g-equivalent of electrolyte is going to be same as1000/c.

Thus, the concept of equivalence conductivity.

$$\Lambda = \kappa \times \frac{1000}{C}$$

$$\Lambda = \kappa \times \frac{1000}{N}$$

Regarding the matter at hand, normality represents N (Hnedkovsky et al., 2005). The equivalence conductivity unit is $ohm^{-1}\ cm^2\ eq^{-1}$.

6.5.3. Molar Conductance

Molar conductor (Figure 6.3) is determined by the conductivity for every ions that are generated through ionization for 1 g-mole for the electrolyte that it dissolved in V ml for solution. Shown by the symbol μ.

Conductivity in Moles

μ =κ× ς (vii)

Assuming a volume for V in mL with 1 g-mole of an electrolyte (Miller, 1996). If the concentration for the solution is given in g-mole each liter.

$$\mu = \kappa \times \frac{1000}{C}$$ $ohm^{-1}\ cm^2\ mol^{-1}$ is unit

$$Equivalent\ conductance = \frac{Molar\ conductance}{n};$$

where,

$$n = \frac{Molecular\ mass}{Equivalent\ mass}$$

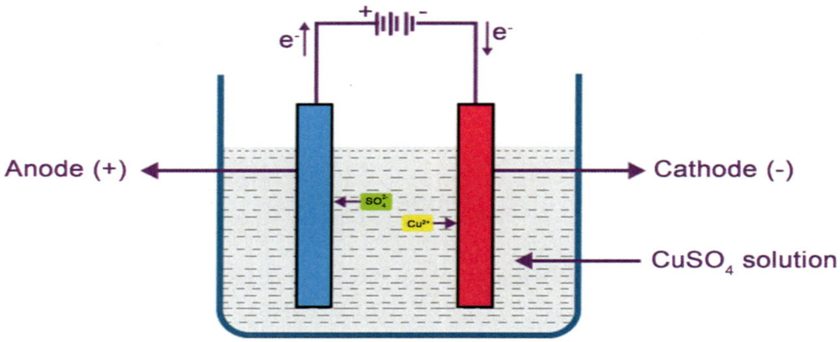

Figure 6.3. *Image of process of molar conductance.*

(Source: Byjus, Creative Commons License).

6.6. KOHLRAUSCH'S LAW

When dissociation completes at infinite dilution, every ion contributes a specific amount to the molar conductivity of the electrolyte. This contribution remains constant regardless of the ion it is linked with or the molecular conductance values ions $\Lambda = \lambda + \lambda\ \lambda C$ or λ. The conductivity of ions for both anion with cation during infinite dilution is represented as λC and λ, accordingly. The ionic conductance is directly related to the ionic motilities. The ionic mobilities of anion, cation shown as u_C and u_a and can be represented by $\lambda_C = ku_C$ and $\lambda_a = ku_a$. The value for k is approximately 96,500 C, which is equivalent to one Faraday.

Tip

Use Kohlrausch's Law to determine the molar conductivity of electrolytes, which helps in understanding their conductive properties more accurately.

Therefore, if one considers that the rise in similar conductance if a solution is diluted is caused by an increase in dissociation of electrolyte, and is clear that electrolyte reaches when it is fully ionized at an extremely high level of dilution. Thus, when the dilution is altered, the equivalent conductance will vary in relation to the extent of dissociation. Therefore,

$$a = \frac{\Lambda}{\Lambda_\infty}$$

Ionic mobility, μ, represents the distance an ion can travel in one second when exposed to one volts per meter.

1. Thus an, $\mu = \lambda^{o} / \Phi$
2. The movement of a single ion is influenced by factors such as its size, solvent viscosity its charge, and temperature.
3. In an aqueous solution, the volume of the aqueous ion is directly related to the charge on the gaseous ion or inversely linked to its size. When a large ion moves within a solution, it encounters increased resistance due to the size that of the solvent particles. Therefore, there is a reduction in both the conductance and ionic mobility. The following list presents the ions in ascending order of their ionic mobilities:

$Li^+ < Na^+ < <K^+ Rb^+ < Cs$; $F^{+ -} < Cl^- < Br^- < I$; $Al^{-\ \ 3+} < Mg^{2+} < Na^+$

4. Interestingly, despite being the smallest ion, the H+ gaseous ion has the highest ionic mobility compared to other ions. A remarkable ionic motion observed within

ions can be attributed towards the exchange of covalent bonds or hydrogen bonds. This allows transfer of charge lacking significant displacement in the ions, as explained by the Grotthus mechanism (Soylu et al., 2017).

Kohlrausch's Law uses

Determination $\ddot{E}_m^0$ of a weak electrolyte: To figure out the Λ_m^0 of an electrolyte that is weak like $CHCOOH_3$, one conduct experimental measurements of the $\Lambda_m{}^0$ values for three electrolytes that are strong.

i. Powerful electrolyte that contains similar cations just like a test electrolyte, such as hydrochloric acid.

ii. Strong electrolyte that contains the similar anion to examine electrolyte, such as $CHCOONa_3$

iii. Powerful electrolytes that contains the same anion as (a) and the cation as (b), specifically NaCl.

1. Determination of degree of dissociation (α):

$$\alpha = \frac{No.\ of\ molecules\ ionised}{Total\ number\ of\ molecules\ dissolved} = \frac{\Lambda_m}{\Lambda_m^0}$$

2. Determination of solubility of sparingly soluble salt:

$$\Lambda_m^0 = \frac{1000\kappa}{C}$$

The molarity for the solution determines the solubility.

3. **Water ionic product determination:** Following Kohlrausch's law, one can calculate molar conductance of water at infinite dilution, denoted as Λm0. This value represents the conductance of a single mole of water if it is fully ionized to produce one mole of H+ ions and one mole of OH⁻ ions that is $\Lambda_m{}^0 (H_2\ O_) = \lambda + \lambda_{H^+}{}^0$ $^0{}_{OH^-}$

Once more, utilizing the following

$$\Lambda_m = \frac{\kappa \times 1000}{C}$$

where C is the molar quantity, or mole L^{-1} or mole dm^{-3}

$$\Rightarrow \quad \Lambda_m = \frac{\kappa}{C},\ \text{where C = concentration in mole } m^{-3}$$

Assuming that Λ_m differs very little from Λ_m^0; $\Lambda_m^0 = \frac{\kappa}{C} \Rightarrow C = \frac{\kappa}{\Lambda_m^0}$

The exact conductance ()κ of clean water is identified through experimental methods (Zou et al., 2021). Afterwards, the level of split water can be calculated via the equation mentioned above. K_w is then calculated as: $K_w = C^2$

6.7. ELECTROCHEMICAL CELLS

6.7.1. Electrolytic Cells

Electrolytic cells have devices that utilize electricity to drive chemical reactions with reduction and oxidation, or to convert electrical energy toward chemical energy (Liang et al., 2020).

6.7.2. Galvanic Cells

This device utilizes a reaction (redox) to transform energy (chemical) into energy (electrical). In other words, it harnesses the power of reactions involving oxidation and reduction to generate electricity. The chemical process causing the production of energy occurs in two distinct compartments. The compartments are made up of an appropriate electrolyte solution along with a metallic conductor. The metallic conductor functions as an electrode, while the compartments housing the electrode along with the electrolyte solution are referred to as half-cells. After the two parts are connected via a salt bridge, electrodes are also connected by a wire with the galvanometer, an electrical current of electricity is initiated. This is a basic form of a voltaic cell (Mancy et al., 1962).

6.7.3. Daniel Cell

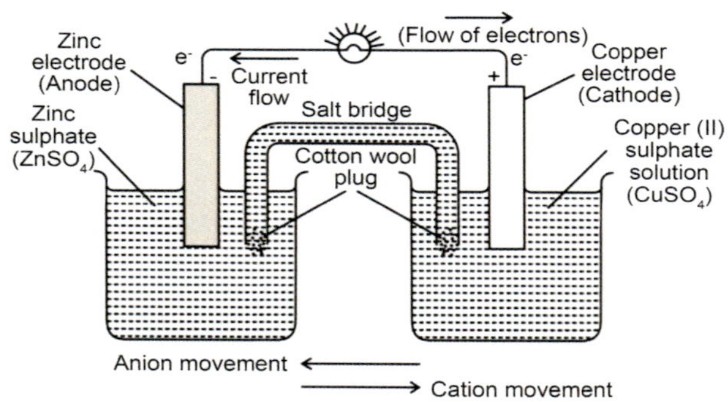

Figure 6.4. *Illustration of Daniel cell.*

(Source: Geeks for Geeks, Creative Commons License).

Oxidation half reaction, $Zn(s) \longrightarrow Zn^{2+}(aq.) + 2e^-$

Reduction half reaction, $Cu^{2+}(aq.) + 2e^- \longrightarrow Cu(s)$

Net reaction $Zn(s) + Cu^{2+}(aq.) \longrightarrow Zn^{2+}(aq.) + Cu(s)$

6.7.3.1. Salt Bridge

Salt bridges typically consist of a U-tube that contains a solution for electrolytes (inert). An electrolyte which is inactive is a type of electrolyte that does not participate in any electrochemical reactions or chemically interact with electrolytes within the two cells (half).

6.7.3.2. Salt Bridge Importance of Salt Bridge

These are uses of a salt bridge:

i. It connects the solutions with two half of a cell and sets circuit of cell.
ii. Inhibits the movement of solutions between the two half-cells.
iii. It maintains electrical neutrality in the solutions of the two half-cells.
iv. It inhibits the formation of liquid-liquid junction possibility, which is a potential disparity that occurs when two solutions meet the other.

The presence of a fractured vertical line and two parallel vertically within a cell reaction signifies a salt bridge existence (Daniel and Errington, 2003).

$Zn \mid Zn^{2+} \parallel Cu^{2+} \mid Cu$

1. **Electrode Potential:** Metal is immersed within a solution containing ions, it acquires a positive or negative charge in relation to the solution. As a result, a clear gradient gets formed among the metal and the solution. Thus, electrode potential is formed by variation in potential. Outcome is contingent upon the characteristics of the electrode, the quantity of ions, and its temperature (Park et al., 2004).
2. **Oxidation Potential:** An electrode has a natural inclination to undergo oxidation, which means it tends to release electrons.

   $M \rightarrow Mn^+ + ne^-$
3. **Reduction potential:** An electrode has a natural inclination to undergo reduction, meaning it readily accepts electrons.

   $Mn^+ + ne^- \rightarrow M$
4. **Standard Electrode Potential:** The normal electrode potential refers to the potential difference that arises in a metal electrode or a solution containing its ions with a molarity of 1 M and a temperature of 25ºC (278 K).
5. The potential magnitude is influenced by several factors:

   i. Characteristics of electrode

ii. The ions amount in solution

iii. Temperature.

6. **Standard Electrode Potential:** Before immersing the electrode in a solution for comparing the potentials between different electrodes, this is important to first determine the amount of ions in that solution or its temperature in the half cell. The voltage differential generated between an electrode of metal or a solution containing its ions at a concentration of 1 M at 2. A typical electrode potential is the potential difference that arises across an electrode of metal and the solution containing ions from it at a concentration of 1 M and at a temperature near 25ºC (298 K). The standard oxidation potential has a negative value.

   5ºC (298 K). The term used to refer to this is standard potential of an electrode. The standard oxidation potential is equal to the negative of the standard reduction potential (Donald et al., 2008).

7. **Cell EMF:** The flow of electrons via the anode towards the cathode is caused by the difference in voltage across both the right and left electrodes.

8. **Standard EMF:** The standard EMF, denoted by, refers to the EMF readings for an electrode beneath typical circumstances (298 K 1, atm) in order with concentrations per unit for ions E^{O}_{cell}

$E_{cell}^{º} = E_{cathode}^{º} - E_{anode}^{º}$ or $E_{cell}^{º} = E_{right\ electrode}^{º} - E_{left\ electrode}^{º}$

6.7.4. Standard Hydrogen Electrode, SHE or NHE (Reference Electrode)

Measuring potential for a single half-cell is not possible, it is feasible to measure the difference in potential between two half-cells. It is important to connect the electrodes to a different electrode that has a known potential. This is known as an electrode of reference, such as the (SHE) standard hydrogen electrode. Such electrode potential is regarded as zero in standard conditions.

The electrode of hydrogen created in this manner constitutes one for the two half-cells among a voltaic cell. A voltaic cell develops by connecting this half-cell with another half-cell. The hydrogen electrode can function as either a cathode or an anode in relation to another electrode (Gross et al., 2016).

6.7.5 Additional Reference Electrodes

i. Calomel Electrode (Figure 6.5)

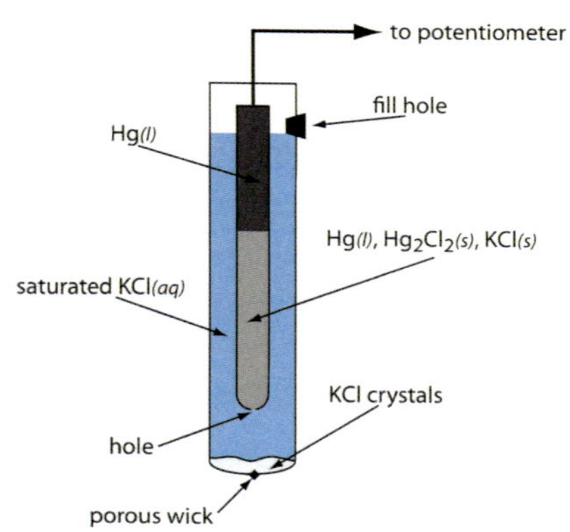

Figure 6.5. *Illustration of Calomel electrode.*

(Source: Topper, Creative Commons License).

Reaction when the electrode functions as a cathode (Grzybowski, 1958):

$$\frac{1}{2}Hg_2Cl_2 + e^- \rightleftharpoons Hg + Cl$$

ii. Electrode made of silver and silver chloride:

The reference electrode is also commonly utilized. The system is both reversible and stable, and it may be built with chloride-containing cells without the need for a liquid connection. A layer of silver chloride is formed by an electrolytic process on a wire made of platinum or silver. Next, the wire is immersed in a solution containing chloride ions. The potential of a typical electrode compared with a common hydrogen electrode is

Electrode shown: 0.2224 V at 298 K. Ag | AgCl | Cl^-

Reaction at the electrode includes: $AgCl + e^- \rightarrow Ag + Cl^-$

6.7.6. Electrochemical Series

6.7.6.1. Characteristics of Electrochemical Series

i. The negative value of SRP shows that when the electrode joins with, SHE, it behaves like an anode and oxidation process happens on it. Hence, a positive charge for SRP suggests the electrode, when connected to SHE, act like a cathode with experiencing reduction.

ii. In the series, materials that have a higher ability to reduce other substances compared to hydrogen are positioned above it.

iii. Substances that have a higher oxidizing power than the H+ ion get positioned at the series next to hydrogen.

iv. The phenomena exhibits a downward trend, with the metals positioned at the top referred to be active metals (Lever, 1990).

6.7.6.2. Application of Electrochemical Series

i. Reactivity of Metals

a. Some metals, particularly alkali metals along with alkaline earth metals, are highly reactive. They react using cool water, easily dissolve in acids with releasing H gas.

b. Metals, such as Sn, Fe, Co, Pb, Ni, cannot react in cool water. However, they do react in steam and release hydrogen gas.

c. Some metals like Cu, Li, Be, Ag, and Au, that are placed below hydrogen, do not react or produce H when in contact with water (Kim et al., 2014).

ii. Electropositive Character of Metals

Metals become less electropositive as one moves down the periodic table.

Displacement reactions

The metal with a lower Standard Reduction Potential (SRP) will replace the metal within its salt solution with a higher SRP value.

Reducing power of metals

The reduction potential of elements diminishes within electrochemical series in a descending order.

Oxidizing nature of non-metals

The oxidizing property of elements rises in a descending order in the electrochemical series.

Thermal stability of metallic oxides

Metal oxide's thermal conductivity decreases as one moves downwards.

Products of electrolysis

The ion that acts as a more powerful oxidizing agent is the one that gets discharged initially at the cathode after deposition (Grotheer et al., 2006).

K, $Ca^{+\ 2+}$, Na, $Mg^{+\ +2}$, Al^{+3}, Zn^{+2}, Fe^{+2}, H, $Cu^{+\ +2}$, Ag, $Au^{+\ +3}$ increasing order

iii. Corrosion of Metals

Corrosion is the process by which a substance deteriorates due to its reaction with the surrounding environment. The corrosion propensity diminishes in a downward direction.

iv. Extraction of Metals

Silver (Ag) and gold (Au) are extracted using the cyanide method.

Two crucial factors that can be deduced from the potential of a cell have the equilibrium constant over the cell reaction along with the shift in free energy over the cell reaction (Jin and Zhang, 2020).

1. Calculating an equilibrium constant using E°_{cell}
2. Calculating the shift in free energy within the standard state E°_{cell}
3. Calculating the non-standard renewable energy change

1. Calculating the Equilibrium Constant using E^{O}_{cell}

Determining the electrochemical cell equilibrium constant, certain information is required. The potential of a cell in its standard state

o The half-reactions that are involved

Figuring out an equilibrium constant involves the use of the Nernst equation.

$$E^{\circ}_{cell} = \frac{RT}{nF} \ln Q$$

When the system is at equilibrium, the value of Q is equal to K. By replacing the values of R, T, and F into the equation, one finds the value of K.

$$E^{\circ}_{cell} = \frac{0.0257}{n} \ln K = \frac{0.0592}{n} \log K$$

2. Determining the Standard State Free Energy Change from E°_{cell}

To calculate the typical state free shift in energy over a cell reaction

1. Find out the $E^{\circ}_{cell.}$
2. Calculate the quantity of electrons moved in the reaction.
3. Calculate the change in Gibbs free energy (ΔG) using the given equation $\Delta Go = -nFE^{\circ}_{cell}$
4. n = number of moles of electrons transferred, ΔG^{o} = standard state free energy change (joules) (Nozik, 1975).
5. E°_{cell} = standard state cell potential (volts or joules/C), F = Faraday's constant (96,485 C/mole e^{-}).

3. Determining the Non-Standard State Free Energy change

To calculate the shift in free energy within a non-standard state:

1. Find cell potential, $E^{\circ}_{cell.}$
2. Calculate the mole number of electrons being transferred n.
3. Determine the Q reaction quotient (Morgantini and Kollman, 1995).
4. Use Nernst equation to figure out the non-standard cell's potential, E°_{cell}. Determine the non-standard free shift in power by utilizing the equation: Δ–G =nFEcell.

6.8. CORROSION

Corrosion refers to the process of a substance weakening and deteriorating due to its interaction with the surrounding environment. Electrochemistry refers to the phenomenon where metals naturally return to the combined state, that is the opposite of a method of extracting metals (Pourbaix, 1974).

6.9. DRY CELL

Within cell after depletion of chemicals, no extra reaction can occur. Regeneration cannot happen by changing the current that flows with cell with a current-driven source for electrical power.

During the operation of the cell, the zinc undergoes oxidation and is converted into Zn2+ ions through the process Zn^{2+}; $Zn \rightarrow Zn^{2+} + 2e^-$ (Anode reaction). Electrons are employed on the (cathode) carbon rod for the ammonium ions reduction.

$2NH^+_4 + 2e^- \rightarrow 2NH_3 + H_2$ (Cathode reaction)

This reaction of cell is $Zn + 2NH^+_4 \rightarrow Zn^{2+} + 2NH_3 + HMnO2$ oxidizes H in the cell. The chemical equation is as follows: $2MnO2 + H_2 \rightarrow 2MnO(OH)$. The ammonia generated at the cathode reacts with the ions of zinc and creates an ion that is complex.

Zn2+ + 4NH3→[Zn(NH3 4)]2+; Ecell is 1.6 volt (Cahoon, 1952).

6.10. FUEL CELL

Fuel cells are an alternative method for converting chemical energy into electrical energy. A primary cell's primary drawback is its limited ability to provide current for a brief duration. This is due to the restricted amount of the oxidizing and reducing agent. However, so long as the external fuel supply is sustained, a fuel cell can continuously generate energy.

An example of such are hydrogen-oxygen fuel cells. The cell consists of three compartments which are separated with a permeable electrode. Hydrogen gas is introduced into one compartment, while oxygen gas is supplied to another compartment. The gasses slowly spread through the electrodes and react with electrodes made of porous carbon, whereas the electrolyte is a resin containing a highly concentrated sodium hydroxide solution in water. The process entails an oxidation of hydrogen on the anode along with the reduction in oxygen on the cathode. Water is generated because of the total cellular response. The occurring reactions include:

Anode: $[H_2(g) + 2OH^-(aq.) \longrightarrow 2H_2O(l) + 2e^-] \times 2$

Cathode: $O_2(g) + 2H_2O(l) + 4e^- \longrightarrow 4OH^-(aq.)$

Overall: $2H_2(g) + O_2(g) \longrightarrow 2H_2O(l)$

These cells are utilized in spacecraft. Fuel cells have high efficiency and are devoid of any pollution. The thermodynamic capacity of fuel cells can be defined as the ratio to the electric free energy produced to the enthalpy for the process (Qiao et al., 2010).

$$\eta = \frac{\Delta G}{\Delta H}$$

6.11. CONCENTRATION CELL

a. **Electrode concentration cells:** The variation in potential is generated in these cells by two similar electrodes with varying concentrations immersed in the same electrolyte solution. An example for this kind of cell is when two hydrogen electrodes are placed in a similar solution of hydrogen ions, but with distinct gas pressure.

When $p_1 > p_{,2}$, oxidation begins at the leftmost electrode while reduction occurs at the right-hand electrode.

$$E_{cell} = \frac{0.0591}{2} \log \frac{(p_1)}{(p_2)} \text{ at } 25^\circ C$$

Within the amalgam cells, identical amalgams consisting of identical metals but with varying concentrations are submerged in a shared electrolytic solution. $M(HgC_1)|M^{n+}|Zn(HgC_2)$

The electromotive force of a cell is determined by the provided expression (Virkar et al., 2000).

$$E_{cell} = \frac{0.0591}{n} \log \frac{C_{2(RHS)}}{C_{1(LHS)}} \text{ at } 25^\circ C$$

b. **Electrolyte concentration cells:** In these cells, each of the electrodes are alike, but are submerged in a solution with varying concentrations of an identical electrolyte. The electrical energy within the cell is generated by the electrolyte's natural tendency to move from a solution with higher concentration to one with lower concentration. Over time, both concentrations happen to balance out. Therefore, initially, the electromotive force of the cell reaches its highest point, and it gradually decreases to zero. This type of cell usually appears in a specific manner: (C2 is greater than C1).

$$M|M^{n+}(C_1)||M^{n+}(C_2)|M \text{ or } \underset{\text{Anode}}{\underline{Zn|Zn^{2+}(C_1)}}||\underset{\text{Cathode}}{\underline{Zn^{2+}(C_2)|Zn}}$$

The expression provided determines the electromotive force (e.m.f) of the cell

$$E_{cell} = \frac{0.0591}{n} \log \frac{C_{2(RHS)}}{C_{1(LHS)}} \text{ at } 25^\circ C$$

Concentration cells are utilized to ascertain the solubility of salts that have low solubility, the valency for each cation within the electrolyte, and the change point between two different forms of a metal utilized for electrodes, among other applications (Yan et al., 2014).

ACTIVITY 6.1.

Construct a Daniel Cell experiment to observe the principles of galvanic cells in action. Measure voltage changes over time and analyze the factors affecting its performance.

SUMMARY

- The chapter begins by introducing essential terminologies in electrochemistry, emphasizing the study of chemical reactions involving electric currents and the interplay of ions. It details the process of electrolysis, highlighting how electrical energy induces non-spontaneous redox reactions, leading to the decomposition of substances into ions.
- Faraday's laws of electrolysis are explored, illustrating quantitative relationships between the amount of substance transformed during electrolysis and the quantity of electricity passed through the electrolyte. Arrhenius' theory of electrolytic dissociation is discussed, outlining how electrolytes dissociate into ions in solution and factors influencing this dissociation.
- The concept of electrical conductance is elaborated upon, distinguishing between conductivity, equivalent conductance, and molar conductance, crucial for understanding ion mobility in solutions. Kohlrausch's law is introduced as a principle governing the conductance of electrolytes.
- Electrochemical cells are explored in depth, differentiating between electrolytic cells, where electrical energy drives non-spontaneous reactions, and galvanic cells, which generate electrical energy from spontaneous redox reactions. The standard hydrogen electrode (SHE) is identified as a reference electrode in electrochemical measurements, while the electrochemical series ranks substances based on their electrode potentials.
- The chapter concludes with discussions on corrosion, including types such as dry corrosion and oxidation corrosion, and methods to mitigate it, like using protective coatings and cathodic protection.

REVIEW QUESTIONS

1. Explain Faraday's laws of electrolysis and their significance in electrochemical reactions.
2. Compare and contrast electrolytic cells and galvanic cells, highlighting their operational differences and applications.
3. Discuss Kohlrausch's law and its relevance in the study of electrolytic solutions.
4. What factors influence the conductivity of electrolytes? Explain with examples.
5. Describe the applications of electrochemical cells in practical scenarios such as in batteries and industrial processes.
6. How does corrosion occur, and what are the different types of corrosion mechanisms? Explain with examples.

MULTIPLE CHOICE QUESTIONS

1. **What is the unit used to measure specific conductance?**
   a. Siemens (S)
   b. Ohms (Ω)
   c. Coulombs (C)
   d. Farads (F)
2. **Which law states that the amount of substance liberated at an electrode during electrolysis is directly proportional to the quantity of electricity passed through the electrolyte?**
   a. Avogadro's law
   b. Faraday's first law
   c. Boyle's law
   d. Charles's law
3. **Which type of corrosion occurs due to the presence of dissimilar metals in contact with each other in an electrolyte?**
   a. Dry corrosion
   b. Oxidation corrosion
   c. Galvanic corrosion
   d. Pitting corrosion
4. **In an electrochemical cell, what is the role of the salt bridge?**
   a. Facilitates electron transfer
   b. Maintains electrical neutrality
   c. Enhances cell voltage
   d. Prevents electrolyte leakage
5. **Which electrode in an electrolytic cell is connected to the negative terminal of the power source?**
   a. Anode
   b. Cathode
   c. Salt bridge
   d. Reference electrode

Answers to Multiple Questions

1. (a) 2. (b) 3. (c) 4. (b) 5. (b)

REFERENCES

1. Arrhenius, S. (1889). IV. Electrolytic dissociation versus hydration. *The London, Edinburgh, and Dublin Philosophical Magazine and Journal of Science, 28*(170), 30–38.
2. Berzins, T., & Delahay, P. (1953). Theory of electrolysis at constant current in unstirred solution. II. Consecutive electrochemical reactions. *Journal of the American Chemical Society, 75*(17), 4205–4213.
3. Bjerrum, N. I. E. L. S. (1909). A new form for the electrolytic dissociation theory. In *Proceedings of the Seventh International Congress of Applied Chemistry Londres* (Vol. 27, pp. 55–50).
4. Cahoon, N. C. (1952). An electrochemical evaluation of manganese dioxide for dry battery use. *Journal of The Electrochemical Society, 99*(9), 343.
5. Cox, M. H., Su, G. W., & Constantz, J. (2007). Heat, chloride, and specific conductance as ground water tracers near streams. *Groundwater, 45*(2), 187–195.
6. Daniel, R. A., & Errington, J. (2003). Control of cell morphogenesis in bacteria: Two distinct ways to make a rod-shaped cell. *Cell, 113*(6), 767–776.
7. Daniell, J. F. (1840). XI. Second letter on the electrolysis of secondary compounds. Addressed to Michael Faraday, Esq. DCLFRS, Fullerian Prof. Chem. Royal Institution, &c. &c. &c. *Philosophical Transactions of the Royal Society of London*, (130), 209–224.
8. Donald, W. A., Leib, R. D., O'Brien, J. T., Bush, M. F., & Williams, E. R. (2008). Absolute standard hydrogen electrode potential measured by reduction of aqueous nanodrops in the gas phase. *Journal of the American Chemical Society, 130*(11), 3371–3381.
9. Garnett, P. J., Garnett, P. J., & Treagust, D. F. (1990). Implications of research on students' understanding of electrochemistry for improving science curricula and classroom practice. *International Journal of Science Education, 12*(2), 147–156.
10. Gewald, T., Lienkamp, M., Lehmkuhl, D., & Hahn, A. (2019). Accelerated aging characterization of lithium-ion cells: Limitation of arrhenius dependency. In *2019 Fourteenth International Conference on Ecological Vehicles and Renewable Energies (EVER), IEEE* (Vol. 2, pp. 1–10).
11. Gross, M. A., Creissen, C. E., Orchard, K. L., & Reisner, E. (2016). Photoelectrochemical hydrogen production in water using a layer-by-layer assembly of a Ru dye and Ni catalyst on NiO. *Chemical Science, 7*(8), 5537–5546.
12. Grotheer, M., Alkire, R., Varjian, R., Srinivasan, V., & Weidner, J. (2006). Industrial electrolysis and electrochemical engineering. *Electrochem. Soc. Interface, 15*(1), 52–54.
13. Grzybowski, A. K. (1958). The standard potential of the calomel electrode and its application in accurate physico-chemical measurements. I. The standard potential. *The Journal of Physical Chemistry, 62*(5), 550–555.
14. Hnedkovsky, L., Wood, R. H., & Balashov, V. N. (2005). Electrical conductances of aqueous Na2SO4, H2SO4, and their mixtures: Limiting equivalent ion conductances, dissociation constants, and speciation to 673 K and 28 MPa. *The Journal of Physical Chemistry B, 109*(18), 9034–9046.

15. Jin, W., & Zhang, Y. (2020). Sustainable electrochemical extraction of metal resources from waste streams: From removal to recovery. *ACS Sustainable Chemistry & Engineering, 8*(12), 4693–4707.

16. Jockusch, R. A., Schnier, P. D., Price, W. D., Strittmatter, E. F., Demirev, P. A., & Williams, E. R. (1997). Effects of charge state on fragmentation pathways, dynamics, and activation energies of ubiquitin ions measured by blackbody infrared radiative dissociation. *Analytical Chemistry, 69*(6), 1119–1126.

17. Kim, E. J., Kim, J. H., Chang, Y. S., Turcio-Ortega, D., & Tratnyek, P. G. (2014). Effects of metal ions on the reactivity and corrosion electrochemistry of Fe/FeS nanoparticles. *Environmental Science & Technology, 48*(7), 4002–4011.

18. Laity, R. W. (1959). General approach to the study of electrical conductance and its relation to mass transport phenomena. *The Journal of Chemical Physics, 30*(3), 682–691.

19. Lever, A. B. P. (1990). Electrochemical parametrization of metal complex redox potentials, using the ruthenium (III)/ruthenium (II) couple to generate a ligand electrochemical series. *Inorganic Chemistry, 29*(6), 1271–1285.

20. Liang, S., Altaf, N. Huang, L., Gao, Y., & Wang, Q. (2020). Electrolytic cell design for electrochemical CO_2 reduction. *Journal of CO_2 Utilization, 35*, 90–105.

21. Mancy, K. H., Okun, D. A., & Reilley, C. N. (1962). A galvanic cell oxygen analyzer. *Journal of Electroanalytical Chemistry (1959), 4*(2), 65–92.

22. Masa, J., Barwe, S., Andronescu, C., & Schuhmann, W. (2019). On the theory of electrolytic dissociation, the greenhouse effect, and activation energy in (electro) catalysis: A tribute to Svante Augustus Arrhenius. *Chemistry–A European Journal, 25*(1), 158–166.

23. Mazloomi, K., Sulaiman, N., Ahmad, S. A., & Yunus, N. A. M. (2013). Analysis of the frequency response of a water electrolysis cell. *International Journal of Electrochemical Science, 8*(3), 3731–3739.

24. Miller, D. G. (1996). Binary mixing approximations and relations between specific conductance, molar conductance, equivalent conductance, and ionar conductance for mixtures. *The Journal of Physical Chemistry, 100*(4), 1220–1226.

25. Morgantini, P. Y., & Kollman, P. A. (1995). Solvation free energies of amides and amines: Disagreement between free energy calculations and experiment. *Journal of the American Chemical Society, 117*(22), 6057–6063.

26. Nozik, A. J. (1975). Photoelectrolysis of water using semiconducting TiO_2 crystals. *Nature, 257*(5525), 383–386.

27. Palit, S. R. (1967). Possibility of a real deviation from Faraday's law of electrolysis. *Indian Journal of Physics, 41*, 309–311.

28. Park, J. H., Kim, D. U., & Paik, S. H. (2004). An analysis of conception types of high school students, chemistry teachers, and pre-service science teachers about salt bridge of Daniel electronic cell. *Journal of the Korean Association for Science Education, 24*(3), 544–555.

29. Pourbaix, M. (1974). Applications of electrochemistry in corrosion science and in practice. *Corrosion Science, 14*(1), 25–82.
30. Qiao, Y., Bao, S. J., & Li, C. M. (2010). Electrocatalysis in microbial fuel cells—from electrode material to direct electrochemistry. *Energy & Environmental Science, 3*(5), 544–553.
31. Schröder, U., Harnisch, F., & Angenent, L. T. (2015). Microbial electrochemistry and technology: Terminology and classification. *Energy & Environmental Science, 8*(2), 513–519.
32. Smolinka, T., Bergmann, H., Garche, J., & Kusnezoff, M. (2022). The history of water electrolysis from its beginnings to the present. In *Electrochemical Power Sources: Fundamentals, Systems, and Applications* (Vol. 2, pp. 83–164). Elsevier.
33. Soylu, M., Coskun, B., Al-Sehemi, A. G., Al-Ghamdi, A. A., & Yakuphanoglu, F. (2017). The validity of Kohlrausch law for the photocurrent transient and the role of N2/Ar flow ratio in photoconductivity of sputtered CoZnO. *Journal of Alloys and Compounds, 712*, 152–163.
34. Thévenot, D. R., Toth, K., Durst, R. A., & Wilson, G. S. (2001). Electrochemical biosensors: Recommended definitions and classification. *Biosensors and Bioelectronics, 16*(1, 2), 121–131.
35. Virkar, A. V., Chen, J., Tanner, C. W., & Kim, J. W. (2000). The role of electrode microstructure on activation and concentration polarizations in solid oxide fuel cells. *Solid State Ionics, 131*(1, 2), 189–198.
36. Wagner, F. T., Lakshmanan, B., & Mathias, M. F. (2010). Electrochemistry and the future of the automobile. *The Journal of Physical Chemistry Letters, 1*(14), 2204–2219.
37. Yan, Y., Timonen, J. V., & Grzybowski, B. A. (2014). A long-lasting concentration cell based on a magnetic electrolyte. *Nature Nanotechnology, 9*(11), 901–906.
38. Zou, Z., Wu, L., Luo, T., Yan, Z., & Wang, X. (2021). Assessment of anion exchange membrane selectivity with ionic membrane conductivity, revised with Manning's theory or the Kohlrausch's law. *Journal of Membrane Science, 635*, 119496.

CHAPTER 7

Corrosion and Its Control

LEARNING OBJECTIVES

After studying this chapter, you will be able to:

1. Identify and describe various types of corrosion, including dry, oxidation, and electrochemical corrosion.
2. Explain the mechanisms and examples of different types of electrochemical corrosion, such as galvanic corrosion, differential aeration corrosion, and stress corrosion.
3. Understand the concept of the Pilling-Bedworth ratio and its significance in corrosion science.
4. Discuss the factors that influence corrosion, focusing on the nature of metals and the characteristics of the corroding environment.
5. Evaluate various corrosion control methods, including design and material selection, cathodic protection, anodic protection, the use of corrosion inhibitors, and protective coatings.
6. Analyze the effectiveness of different corrosion control strategies in preventing or mitigating corrosion in various applications.
7. Apply knowledge of corrosion control to practical scenarios in industrial and environmental settings.

KEY TERMS FROM THIS CHAPTER

Dry corrosion
Galvanic corrosion
Stress corrosion
Corrosion inhibitors
Electrochemical corrosion
Differential aeration corrosion
Pilling-Bedworth ratio

INTRODUCTION

Corrosion refers to the progressive disintegration, corrosion, or deterioration of metal or alloys caused by chemical or electrochemical reactions with the surrounding environment. It is a highly significant issue encountered by advanced industrial societies. Approximately 30–40% of the steel and iron produced each year is utilized to replace corroded items (Ferrando, 1989).

Metals are generally available in nature in their compound forms (ores). These are the stable forms of metals. However, for getting useful engineering materials, that is the metallic form, a large amount of energy is needed to convert ores to their metallic form. So metals are comparatively unstable and have a general tendency to revert to their compound form. This process is known as corrosion (Figure 7.1) (Loto, 2017).

Example: Rusting of Iron

$$2\ Fe(s) + \ 3/2\ O_2(g) + x\ H_2O \rightarrow Fe_2O_3.xH_2O(s)$$

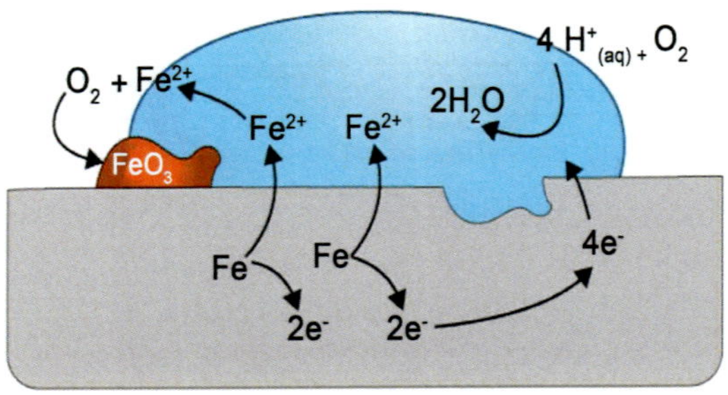

Figure 7.1. *Example of corrosion reaction.*

(Source: Byju's, creative Commons License).

The following are the effects of corrosion:

(i) Loss of metal leads to a reduction in thickness which may cause a reduction in mechanical strength leading to sudden structural collapse or breakdown (Amin and Ibrahim, 2011).

(ii) Mechanical degradation of valves, pumps, and so forth, or obstruction of pipes due to the accumulation of solid corrosion byproducts.

(iii) Time is lost in making available the parts to replace the damaged parts or equipment.

(iv) The fluids stored in pipes or storage tanks get contaminated due to chemical reactions (Schock, 1989).

7.1. TYPES OF CORROSION

There are two prevalent forms of corrosion.

7.1.1. Dry Corrosion (Chemical Corrosion)

Dry corrosion, also known as chemical corrosion, refers to the process in which air gases directly attack a material with no moisture, which acts as a conducting medium (Hœrlé et al., 2004). The atmospheric gases consist of oxygen, halogens, oxides of sulfur, nitrogen, hydrogen sulfides, and so on. There exist three distinct forms of dry rusting (Zhu et al., 2000).

7.1.2. Oxidation Corrosion

It happens when oxygen directly attacks metals at either low or high temperatures, typically in a lack of moisture. Alkali metals (such as Na, Li, K, Rb, etc.) and metals from alkaline earths (such as Be, Ca, Sr, etc.) undergo fast oxidation at low temperatures. Nevertheless, metals other than silver (Ag), gold (Au), and platinum (Pt) undergo oxidation when exposed to high temperatures (Yang et al., 2007).

The process of oxidation initiates at the metal's surface, resulting in the formation of metal oxides, as indicated by the following reaction:

$4\ M \quad 4\ M^{n+} + 4n\ e$ (oxidation) $O_2 + 4n\ e \quad n\ O^{2-}$ (reduction)

The overall reaction is: $4\ M + n\ O_2 \quad 2\ M_2O_n$

Further process of corrosion continues either due to the diffusion of oxide ions into the inner metal layer or metal ions diffusing out to the surface (Singh et al., 2007) (Figure 7.2).

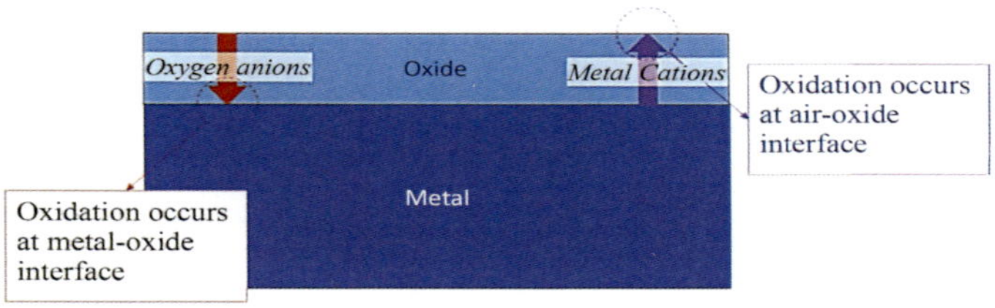

Figure 7.2. *The mechanism of dry corrosion in metal oxides.*

(Source: Lumen Learning, Creative Commons License).

Nevertheless, the composition of the metal oxide layer created at the boundary is crucial in determining the degree of subsequent metal corrosion. The metal oxides that can be generated are classified into the following types:

Remember

Remember that corrosion refers to the gradual destruction of materials by chemical or electrochemical reaction with their environment.

1. Stable metal oxides: These oxides are resistant and create a barrier that shields metals from further oxidation. Examples include oxide layers of aluminum (Al), lead (Pb), copper (Cu), tin (Sn), and other similar materials.
2. Unstable metal oxides: This category of oxide layer undergoes decomposition into metal and oxygen immediately after its formation. No signs of oxidation corrosion are detected for the metals. Examples: Silver and gold oxides.
3. Volatile metal oxides: The metal oxide film evaporates immediately upon formation, exposing the underlying metal to further environmental corrosion. Therefore, fast rusting occurs. Example: MoO_3 (Mahesh and Raman, 2014).

Porous metal oxides are characterized by the presence of pores and fissures in their oxide layers, which facilitate the penetration of ambient oxygen into the underlying metal layers. This leads to unhindered and quick corrosion of the metal. The process persists until the entire metal is transformed into its corresponding oxides. Examples: Lithium oxides, Magnesium oxides, Sodium oxides, etc. (Honey et al., 1986).

Did you know?

Did you know that liquid metal corrosion occurs due to the dissolution of a metal in liquid metal alloys, often accelerated by high temperatures and reactive environments?

7.1.3. Corrosion by other gases, such as Cl_2, SO_2, H_2S, and NO_x

These gases undergo chemical reactions with the metal, resulting in the formation of corrosion products. These products can be either protective or non-protective, depending on the chemical environment. When dry chlorine comes into contact with silver, it undergoes a reaction to produce silver chloride (AgCl), which forms a protective layer. Conversely, tin chloride ($SnCl_4$) is both volatile and porous. Within the petroleum industry, steel is susceptible to attack by H_2S at elevated temperatures, resulting in the formation of a porous FeS scale (Pan et al., 2005).

7.1.4. Liquid Metal Corrosion

This form of corrosion arises from the chemical reaction between liquid metals, flowing at elevated temperatures, and solid metal or alloy surfaces (Figure 7.3). Corrosion occurs when molten metal flows through metallic pipes in various sectors, leading to dissolution or interior penetration. As an illustration, the process of dissolving most metals involves liquid mercury-producing amalgams, which subsequently leads to their corrosion. It is also present in devices utilized for the cooling system of nuclear power plants. The presence of coolant, specifically liquid sodium metal, results in the corrosion of cadmium within the nuclear plant (Cui et al., 2018).

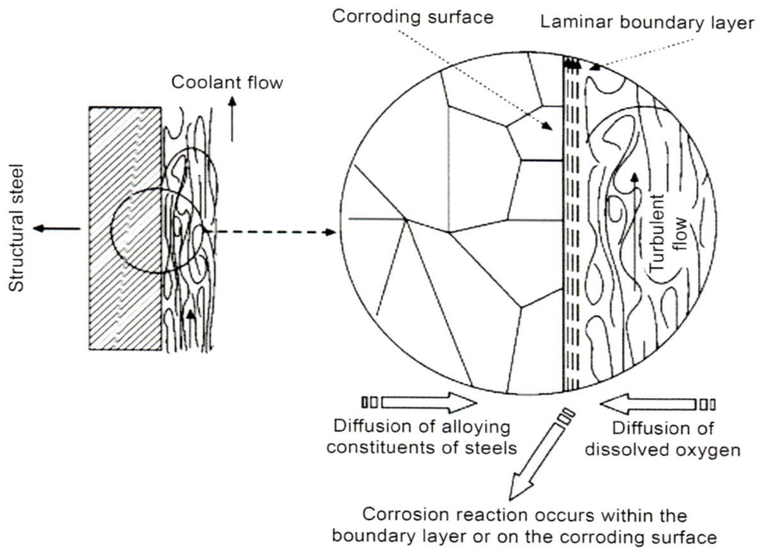

Figure 7.3. *Illustration of liquid metal corrosion.*

(Source: Baldev Raj, Creative Commons License).

7.1.5. Wet or Electrochemical Corrosion

Wet corrosion refers to the process of corrosion that takes place when a metal or two different metals or alloys are partially or fully immersed in a conducting liquid, often an aqueous solution. Wet corrosion occurs when anodic and cathodic areas form, allowing corrosion current to pass through a conductive liquid or solution. Corrosion takes place in the anodic region as a result of oxidation, which involves the breakdown of metals and the release of electrons (Saurbier et al., 1994).

$$M \longrightarrow M^{n+} + n\ e$$

The cathodic process utilizes these electrons, and its occurrence is influenced by the components present in the corrosion medium. The corrosive medium can be generically categorized into four types.

When the medium is supplied with air and is close to being neutral, the oxygen undergoes a reduction in the presence of water (H_2O) to form OH- ions.

$$O2 + 2\ H2O + 4\ e- \longrightarrow 4\ OH^-$$

When the medium contains air and is acidic, the oxygen undergoes reduction to produce water (H_2O).

$$O2 + 4\ H+ + 4\ e- \longrightarrow 2\ H_2O$$

When the medium is de-aerated and nearly neutral, the hydrogen is released along with OH- ions.

$$2\ H2O + 2\ e^- \longrightarrow H_2 + 2\ OH^-$$

When the medium is free of air and has an acidic pH, hydrogen gas is released (Gong et al., 2020).

$$2H^+ + 2\ e^- \longrightarrow H$$

Example: Rusting of Iron in an aerated and neutral medium

Corrosion occurs when iron comes into contact with an aerated and neutral media, leading to the following chemical processes:

Anodic reaction: $Fe \rightarrow Fe^{2+} + 2\ e^-$

Cathodic reaction: $\frac{1}{2}\ O_2 + H_2O + 2\ e^- \rightarrow 2\ OH^-$

The metal ions generated at the anode react with hydroxide ions at the cathode because of the high mobility of Fe^{2+} ions, resulting in the formation of $Fe(OH)_2$. The ferrous hydroxide undergoes additional oxidation to become hydrated ferric oxide, often known as rust (Huang et al., 2013).

$$Fe^{2+} + 2\ OH^- \rightarrow Fe(OH)_2$$

$$Fe(OH)_2 + \frac{1}{2}\ O_2 + (x-2)\ H_2O \rightarrow Fe_2O_3 \cdot xH_2O$$

In conditions of oxygen scarcity, the resulting corrosion product is likely to be Fe_3O_4.

$$Fe(OH)2 + \frac{1}{2}\ O2 \longrightarrow Fe_3O_4 \cdot 3H_2O$$

7.2. TYPES OF ELECTROCHEMICAL CORROSION

7.2.1. Differential Metal Corrosion (Galvanic Corrosion)

Galvanic corrosion (Figure 7.4) happens when two metals with different properties come into touch with one another in a corrosive environment. Metals with a low reduction potential experience oxidation, which is commonly known as corrosion. The corrosion rate increases in direct proportion to the magnitude of the electrodes potential difference across the anode and cathode (Munn et al., 1991).

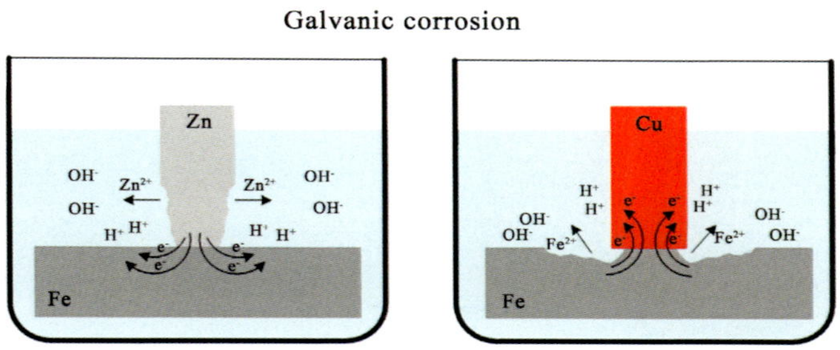

Figure 7.4. *Figure of galvanic corrosion.*

(Source: Subs tech, Creative Commons License).

Examples of galvanic corrosion include:

The above information provides the standard reduction electrode potentials of various metals.

The reduction potentials of the following metals are as follows:

Mg (–2.52 V), Fe (–0.44 V), Zn (–0.76 V), Cu (0.34 V), Ag (0.80 V).

Determine the metal that experiences corrosion in the bimetallic couples: The combinations of elements mentioned are as follows: iron and copper (Fe/Cu), zinc, and iron (Zn/Fe), steel, and copper (Steel/Cu), zinc, and silver (Zn/Ag), and zinc and copper (Zn/Cu). Precautionary measures: By interposing a slender insulating layer between two metallic surfaces. By choosing materials with similar electrode potentials (i.e., minimal electrode potential difference) (Doig and Flewitt, 1979).

7.2.2. Differential Aeration Corrosion

It happens when a metal is exposed to varying levels of air or oxygen content. Due to the oxygen requirement of the cathodic reaction, the region of the metal that is in contact with a larger concentration of oxygen functions as the cathode. Conversely, the metallic portion that is exposed to a lower quantity of oxygen functions as the anodic zone, leading to corrosion in that specific area (Miyasaka et al., 1991).

Some examples of differential aeration corrosions are described below.

7.2.2.1. Water-Line Corrosion

This type of corrosion occurs in maritime vessels, water storage steel tanks, or steel pipelines that carry liquid and are exposed to the atmosphere. The metallic section located near the waterline has a higher anodic potential compared to the section situated above the waterline. Consequently, the metal located directly below the water's surface experiences corrosion. The responses are:

Anodic reaction: $Fe \rightarrow Fe^{2+} + 2\ e$

Cathodic reaction: $\frac{1}{2}\ O_2 + H_2O + 2\ e- \rightarrow 2\ OH^-$

$Fe^{2+} + 2\ OH^- \quad Fe(OH)_2$

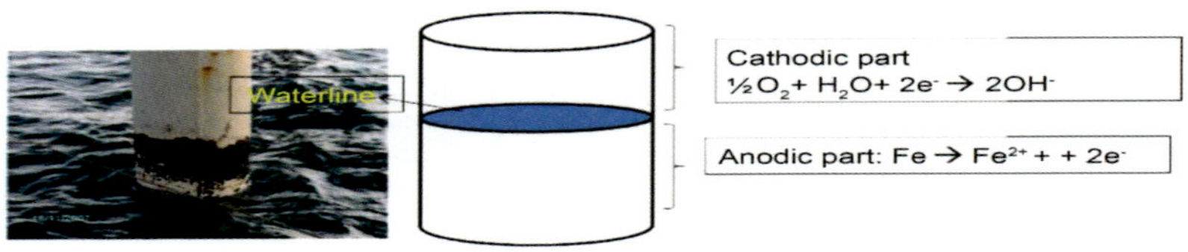

Figure 7.5. *Image showing waterline corrosion.*

(Source: Lumen Learning, Creative Commons License).

Marine vessels are subject to corrosion, whereas ships that have been submerged in seawater for an extended period do not experience corrosion. When a ship sinks, it is submerged in water and is exposed to a consistent amount of air. As a result, the ship does not experience varying levels of corrosion (Zhang et al., 2023).

7.2.2.2. Pitting Corrosion

It entails a targeted attack that leads to the creation of cavities. This corrosion arises when a minute section of the metal surface is flawed (with cracks) or occupied by dust, scale, sand, or a water droplet. The metal behind the water droplet or dust deposit is subjected to reduced levels of oxygen and experiences corrosion. After the formation of a pit, the rate of corrosion escalates. Accelerated corrosion occurs beneath the deposit as a result of a small anodic area and a large cathodic area. Preventive measures include ensuring surfaces are prepared with the highest quality finish, thoroughly eliminating any pollutants from the surface, and designing and manufacturing in a way that prevents the accumulation of liquid on the surface (Casillas et al., 1994).

7.2.2.3. Crevice Corrosion

It typically happens at the junctions where two metallic items, such as bolts, nuts, rivets, and so forth, come into touch with liquids. The crevice area experiences oxygen deficiency, causing it to become an anodic region compared to the rest of the metal, which serves as the cathode. Crevice corrosion is the phenomenon when two metals in contact with each other experience rapid corrosion when exposed to a corrosive environment (Figure 7.6).

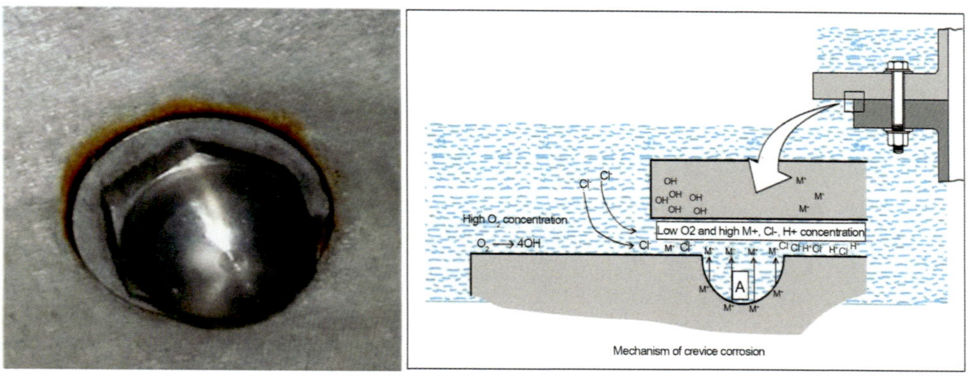

Figure 7.6. The *image and mechanism of crevice corrosion.*

(Source: Hu et al., 2011, Creative Commons License).

7.2.3. Stress Corrosion

Pitting corrosion is a form of localized electrochemical corrosion that occurs when a metal is exposed to both static tensile tension and a corrosive environment. Metal components undergo non-uniform stress during their production processes, which include cold working, bending, quenching, rolling, pressing, and heat treatment. Stressed metal atoms consistently exhibit greater energy levels than unstressed ones. Consequently, the areas under stress exhibit higher reactivity and function as the anode, while other regions serve as the cathode. This leads to corrosion specifically in the stressed areas in a corrosive environment.

7.2.3.1. Seasonal Cracking (Stress Corrosion of Brass)

This type of corrosion was first observed during the British rule in India during monsoon season. Due to diminished military activity during monsoon season, brass cartridges are stored for later usage. After some days, it was found that cracks had developed on the surface of the cartridges. It takes place due to the action of susceptible agents like Cu and its alloys and attacking agents like ammonia or amine and the residual stress (Anderson and Grew, 1977).

7.2.3.2. Caustic Embrittlement of Steel Boiler

Corrosion takes place in mild steel when it is exposed to an alkaline solution under conditions of high temperature and stress. This phenomenon is commonly linked to steam boilers and heat transfer equipment, where mild steel is corroded by highly alkaline water at the vulnerable areas of the boiler when the operating pressure ranges from 10 to 20 atmospheres. Microscopic fissures may form at the strained section of the boiler, allowing alkaline pollutants to infiltrate by capillary action. The water undergoes evaporation, resulting in the accumulation of caustic soda in the crevices, so intensifying its concentration. When iron is subjected to stress, it functions as an anode and undergoes corrosion, which ultimately leads to the collapse of the boiler.

The boiler water contains sodium carbonate (Na_2CO_3), which undergoes hydrolysis at high temperatures to produce sodium hydroxide (NaOH).

$$Na2CO3 + H2O \longrightarrow heat\ 2\ NaOH + CO2$$

The NaOH solution infiltrates the small fractures and gaps, where it chemically interacts with the iron to produce sodium ferroate (Na_2FeO_2). The sodium ferroate undergoes thermal decomposition, resulting in the formation of NaOH. The NaOH then combines with iron, leading to the occurrence of corrosion.

$$2\ NaOH + Fe + heat \rightarrow Na_2FeO_2 + H_{2.}$$

$$3\ Na_2FeO_2 + 4\ H_2O\quad heat \rightarrow Fe_3O_4 + 6\ NaOH + 4\ H_{2.}$$

Preventive steps can be taken to mitigate the risk of corrosion. These procedures include the application of protective coatings, the use of corrosion inhibitors, and the implementation of stress-releasing heat treatment (Soltanloo et al., 2022) (Table 7.1).

Table 7.1. *Tabular Representation of Difference Between Dry Corrosion and Wet Corrosion*

(Source: Doig, P., Creative Commons License).

Dry Corrosion	Wet corrosion
1. It takes place in the absence of moisture	1. It takes place in the presence of moisture
2. It involves a direct attack of atmospheric gases.	2. It involves the development of a large number of tiny galvanic cells.
3. It is less prevalent	3. It is more prevalent
4. It can be explained by chemical reactions	4. It can be explained by electrochemical theory.
5. It is comparatively a slow process	5. It is a fast process
6. This corrosion is uniform.	6. This corrosion is non-uniform as corrosion takes place only at the anodic region, not at the cathodic part.

7.3. PILLING-BEDWORTH RATIO

When metals come into contact with the environment, they undergo oxidation, and a coating of oxide is created. If the oxide is more chemically stable than the metal, it acts as a protective barrier against further corrosion (oxidation) of the metal. In 1923, N. B. Pilling and R. E. Bedworth suggested that the metal's resistance to additional corrosion is dependent on how well the oxide layer sticks to the metal surface. To achieve effective oxidation resistance, it is crucial for the oxide layer to tightly attach to the metal surface. The degree to which the oxide adheres to the metal surface is quantified by the Pilling-Bedworth Ratio or Specific Volume Ratio (R). This ratio can be determined using the following calculation:

Pilling-Bedworth Ratio or Specific volume ratio (R) = (volume of metal oxide) / (volume of metal from which oxide is formed)

$$= \frac{M_{oxide}}{d_{oxide}} \Big/ n\left(\frac{A_{metal}}{d_{metal}}\right)$$

Where, Moxide represents the molecular weight of the metal oxide and Ametal represents the atomic weight of the metal. The terms "doxide" and "d metal" represent the density of the metal oxide and the metal, respectively. The variable "n" denotes the number of metal atoms present in each molecule of the metal oxide.

This rule outlines the characteristics of the oxide layer that forms during corrosion, specifically its protective and non-protective properties.

1. If $R < 1$, the oxide layer is non-protective. Examples: oxides of Mg, Na, K, Ca.
2. If $1 < R < 2$, the oxide layer is passive and causes the corrosion process to slow. Examples: Oxides of Fe, Ni
3. If $R > 2$, a stable oxide layer is formed and further corrosion is prevented. Examples: Oxides of Cr, Si, Nb (Arurault, 2008).

7.4. FACTORS AFFECTING CORROSION

The rate of corrosion of two iron rods will vary if they are exposed to various environmental conditions, such as one being immersed in purified water and the other in saline water. Likewise, when an iron rod and a zinc rod are exposed to the same corrosive environment, there will be a disparity in their corrosion rates. Hence, the corrosion rate is contingent upon the characteristics of the metals and the environment (Glass et al., 1991).

7.4.1. Nature of Metals

1. **Position in the Galvanic Series:** Metals having low reduction electrode potentials are highly reactive and therefore more prone to corrosion. Therefore, as the reduction potential increases, the propensity of a metal to corrode diminishes. Metals such as lithium (Li), sodium (Na), magnesium (Mg), and zinc (Zn) experience significant corrosion in corrosive environments due to their low reduction potentials. Conversely, noble metals such as Ag, Au, Pt, and so forth have strong reduction potentials and are hence less prone to corrosion (Stack et al., 2010).

7.4.1.1. Electrode Potential Difference

The difference in electric potential between the anodic and cathodic regions is referred to as the open circuit potential difference (OCPD). The higher the OCPD, the greater the rate of corrosion. For instance, the electrical potential difference between iron (Fe) and silver (Ag) is 1.14 volts, while the potential difference between iron (Fe) and tin (Sn) is 0.84 volts. Consequently, iron corrodes more rapidly when exposed to silver than when exposed to tin in a corrosive environment.

7.4.1.2. Surface State of the Metal

The rate of the corrosion process is also influenced by the state or circumstances of the metal surface. For instance, when the surface is coated with grit or dust particles, it leads to the occurrence of intense pitting corrosion. The existence of water droplets also enhances the corrosion process at the surface. The presence of heterogeneity on the metal surface contributes to an elevated corrosion rate. The presence of unequal tension contributes to increased corrosion rates (Landolfo et al., 2010).

7.4.1.3. Hydrogen Overvoltage

During the majority of electrochemical corrosion processes, the cathodic reaction that competes involves the release of hydrogen gas as a result of the reduction of hydrogen ions. Hydrogen overvoltage refers to the propensity of an electrode to release hydrogen gas. A metal that has little hydrogen overvoltage is susceptible to corrosion. Low hydrogen overvoltage enhances the speed of the cathodic reaction, leading to an increase in the rate of the anodic reaction and thus, the corrosion rate.

7.4.1.4. Formation of Protective Film of Corrosion Products by Metals

Some metals have a propensity to form passive, yet protective coatings when exposed to corrosive substances. The passive layers, which are stable, highly insoluble, and have poor conductivity, serve as a protective barrier, preventing more corrosion. Metals such as aluminum (Al), titanium (Ti), and chromium (Cr) form a protective layer on their surface, rendering them resistant to corrosion. However, if the layer of product of corrosion is soluble, non-uniform, volatile, or porous, then corrosion will persist. For instance, the presence of metal oxide layers composed of iron (Fe) and zinc (Zn) is insufficient to fully inhibit the corrosion process due to their porous nature (McMahon et al., 2019).

7.4.1.5. Relative Area of Anode and Cathode

When the anode has a smaller surface area compared to the cathode, the corrosion rate increases. The reason is attributed to the elevated anodic current density. For instance, every opening in the tin-coated iron leads to significant corrosion because of the tiny anodic area and the huge cathodic area.

7.4.2. Nature of Corroding Environment

7.4.2.1. pH

The corrosion rate increases as the pH decreases. Typically, the corrosion rate is higher in an acidic environment. Severe corrosion occurs when the pH is less than 3, even without the presence of air, since hydrogen gas is released at the cathodic region. When the pH level is over 10, the corrosion of iron (Fe) almost completely stops because a protective layer of hydrous oxide forms. Nevertheless, metals such as aluminum (Al) and zinc (Zn) experience significant corrosion when exposed to strongly alkaline solutions (Wang et al., 2015).

7.4.2.2. Temperature

The speed of any chemical reaction often correlates positively with temperature, and higher temperatures also augment the electrical conductivity of the media. Corrosion occurs by either chemical or electrochemical reactions, and its pace is directly proportional to temperature. Increasing the temperature leads to a proportional increase in the corrosion rate.

7.4.2.3. Conductance of the Medium

Electrochemical corrosion is characterized by the transport of electrons. Conducting media enhances electron transmission, hence accelerating corrosion. Consequently, the corrosion rate is elevated in saline water compared to regular water.

7.4.2.4. Humidity

Humidity is the quantification of the amount of water vapor present in the air. The rate of corrosion escalates as humidity rises until it reaches a specific threshold known as critical humidity. However, the rate of corrosion experiences a sudden and significant rise if the critical humidity threshold is surpassed (Jiang et al., 2014).

7.5. CORROSION CONTROL METHODS

Corrosion control is a more feasible approach than corrosion avoidance, as the corrosion of a metal is an inherent process that cannot be entirely halted, although its rate can be managed. Various techniques can be employed to manage the corrosion process.

7.5.1. Design and Materials Selection

The complexity of surface treatments increases as the number of angles, edges, corners, and internal surfaces in a design increases. Therefore, it is advisable to minimize the use of L-, T-, and U-shaped profiles in construction to the greatest extent possible.

Given the rapid corrosion rate in the presence of moisture, it is imperative to design the structure in a way that minimizes moisture retention.

To prevent differential metal corrosion, it is advisable to use a single metal instead of using dissimilar metals. If it is necessary to employ them, metals with a low difference in electrode potential should be utilized. A non-conductive material should be placed between the two metals with different properties.

To minimize corrosion, it is advisable to avoid using a smaller anode and a larger cathode, as the rate of corrosion is influenced by their relative sizes (Elliott, 2003).

7.5.2. Cathodic Protection

The theory of cathodic protection is transforming the entire metal into a cathodic site, preventing corrosion at the anode while leaving the cathode undamaged. This can be accomplished by two approaches as outlined below:

7.5.2.1. Sacrificial Anode Method

In this context. the metal structure that needs protection is linked to a more reactive metal, which serves as an auxiliary anode, while the base metal functions as the cathode. Consequently, the primary metal is shielded while the secondary metal experiences corrosion. The sacrificial anodic approach is named as such because it involves sacrificing anodic metals to safeguard the metal structure. Whenever necessary, new auxiliary anodes are used to replace the depleted sacrificial anodes. Zinc (Zn), magnesium (Mg), and aluminum (Al) are the sacrificial anodes that are most frequently utilized. For instance, to safeguard the subterranean pipelines, a magnesium block is attached to them. Magnesium bars are affixed to the sides of ocean-going vessels to manage and prevent corrosion (Park and Kim, 2020).

7.5.2.2. Impressed Current Method

This technique involves safeguarding the underlying metal by administering a direct current (DC) that exceeds the current responsible for corrosion. This is achieved by establishing a connection between the negative terminal of a direct current (DC) source and the base metal, while the positive terminal is connected to an inert anode, such as graphite. The base metal, when serving as the cathode, remains unaffected by corrosion.

7.5.3. Anodic Protection

The underlying idea of this technology is to induce the metal structure to exhibit greater anodic behavior by applying an impressed anodic current, hence preventing corrosion. Metals such as titanium (Ti), chromium (Cr), nickel (Ni), and alloys like steel, can exhibit passivity when used as anodes by producing oxide layers. In such instances, the administration of an appropriate anodic current renders them inactive and reduces their rate of breakdown. The potential needed to safeguard the metal can be derived from the potential-current graph.

Tip

Anodic protection involves applying an external electrical current to ensure a metal surface remains in a passive state, minimizing corrosion.

One benefit of this approach is that it necessitates a low current. Nevertheless, it is not possible to eliminate the corrosion rate, as can be achieved with cathodic protection. One limitation of this approach is that it may not be used on metals that do not undergo passivation. This approach is employed in the transportation of very concentrated acids (Cowley et al., 1968).

7.5.4. Corrosion Inhibitors

Corrosion inhibitors are chemicals that hinder either the anodic or cathodic reactions, hence reducing corrosion. There are two categories of inhibitors.

7.5.4.1. Anodic Inhibitors

These inhibitors hinder the creation of metal ions (anodic reactions) and hence slow down the process of corrosion. This is accomplished by introducing anions, such as chromates, molybdates, tungstate, and so forth, which react with the metal ions produced at the anode. As a consequence, the creation of salts with low solubility occurs, which then accumulate on the anodic site and create a protective barrier between the metal and the corrosive substance. Therefore, the occurrence of more corrosion is inhibited (Cavallaro et al., 1964).

7.5.4.2. Cathodic Inhibitors

These substances work by preventing the cathodic process that releases hydrogen gas in acidic solutions and hydroxide ions in alkaline environments. Organic cathodic inhibitors, such as amines, mercaptans, thioureas, and sulfoxides, create a protective layer on the cathodic region. This layer prevents the formation of hydrogen gas and hence reduces the rate of corrosion.

Inorganic cathodic inhibitors, such as magnesium sulfate, manganese sulfate, nickel sulfate, and zinc sulfate, are utilized in a neutral or alkaline environment. These inhibitors chemically react with hydroxide ions (OH-) that are released at the cathode. Insoluble hydroxides are formed, which then generate a protective coating on the cathodic region, thus preventing corrosion (Nam et al., 2018).

7.5.5. Protective Coating

Protective coatings are applied to surfaces to inhibit corrosion. The coatings might be either metallic or organic. The term "inorganic coatings" refers to coatings that are not composed of organic materials.

7.5.5.1. Metallic Coatings

There are two types of metallic coatings: anodic and cathodic.

1. Anodic metal coatings are created by applying a layer of metals that are more reactive, have higher oxidation potentials, or have lower reduction potentials onto a base metal. These coatings have a higher anodic potential than the base metal. For instance, metals such as aluminum (Al), zinc (Zn), magnesium (Mg), and cadmium (Cd) can be applied as a coating over iron.
2. Cathodic metal coatings are created by applying noble metals or metals with high reduction potentials, or metals with low oxidation potentials, onto base metals. Cathodic coatings demonstrate efficacy just when they exhibit continuity, impermeability, and absence of gaps.

Metallic coatings can be applied using many techniques, including cladding, hot dipping, and metal spraying.

1. Cladding refers to the use of sheet-like materials to protect metal surfaces. The metal sheet that requires protection is placed between two sheets of anodic metals and compressed by a rolling process involving pressure and heat. Example: Application of duralumin sheets to cover and protect aluminum.
2. Hot dipping is a process where the base metal is submerged in a molten form of metal.

3. Galvanization refers to the application of a zinc coating on iron.
4. Tinning refers to the process of applying a layer of tin onto iron.
5. Metal spraying is accomplished by the flame spraying process, in which metal particles are propelled onto the surface of the metal to be protected using an air blast (Priyantha et al., 2003).

7.5.5.2. Organic Coatings

These coatings consist of ingredients that are applied to safeguard or enhance the appearance of a surface. These encompass paints, varnishes, and industrial maintenance coatings that can be utilized on both stationary and mobile sources. The application of organic paints to metal surfaces is the most commonly employed method for preventing corrosion. Organic paint is a mixture of finely divided particles in a medium that contains non-volatile substances, such as drying oils and resins, which can form a film (Nazari et al., 2022).

7.5.5.3. Inorganic Coatings (Ceramic Coatings)

Thin film coatings can be effectively achieved using hard materials such as carbides, nitrides, borides, and silicide's belonging to Groups 14, 15, and 16 of the periodic table. These coatings are created by incorporating nitrogen, hydrocarbons, or silicide's into the sputtering process, which involves depositing a metal onto the surface of another material using high-energy ions (Shao et al., 2015).

ACTIVITY 7.1.

Investigate stress corrosion by setting up experiments with metal samples under mechanical stress in corrosive environments to observe the effects on corrosion rates.

SUMMARY

- The chapter begins by categorizing corrosion into different types, including dry corrosion, oxidation corrosion by gases (such as Cl_2, SO_2, H_2S, and NOx), liquid metal corrosion, and wet or electrochemical corrosion. Each type is examined in terms of their underlying chemical processes and environmental conditions that contribute to material degradation.
- Electrochemical corrosion, a prominent type discussed, encompasses phenomena like galvanic corrosion, differential aeration corrosion, and stress corrosion, each characterized by specific conditions that accelerate material deterioration.
- The Pilling-Bedworth ratio is introduced as a critical parameter in determining whether a metal will corrode or not under certain conditions, providing insights into the stability of metal oxides relative to the metal itself.
- Factors influencing corrosion are thoroughly explored, including the nature of metals, environmental conditions (such as pH, temperature, humidity), and the presence of contaminants. Understanding these factors is crucial for implementing effective corrosion control strategies.
- The chapter then transitions into various methods for corrosion control, emphasizing proactive measures like design and material selection, cathodic, and anodic protection techniques, corrosion inhibitors, and protective coatings. These methods aim to minimize or prevent corrosion by altering the corrosive environment or protecting the material surface from it.

REVIEW QUESTIONS

1. Differentiate between dry corrosion and wet (electrochemical) corrosion mechanisms. Provide examples of each.
2. Explain the factors influencing corrosion and how they affect the degradation of metals in various environments.
3. Discuss the methods of corrosion control, including cathodic protection, anodic protection, and the use of corrosion inhibitors.
4. What is stress corrosion cracking (SCC)? Describe its causes and preventive measures.
5. Explain the Pilling-Bedworth ratio and its significance in the prevention of metal corrosion.
6. How does differential aeration corrosion occur? Provide examples of industries where this type of corrosion is particularly problematic.
7. Discuss the role of the standard hydrogen electrode (SHE) in electrochemical measurements and its importance in the study of corrosion.

MULTIPLE CHOICE QUESTIONS

1. **What type of corrosion occurs due to the action of chlorine gas on metals?**
   a. Dry corrosion
   b. Oxidation corrosion
   c. Pitting corrosion
   d. Corrosion by Cl_2
2. **Which factor affects the rate of corrosion based on the properties of metals?**
   a. Temperature
   b. pH of the environment
   c. Humidity
   d. All of the above
3. **What is the purpose of cathodic protection in corrosion prevention?**
   a. To enhance metal conductivity
   b. To reduce oxygen concentration
   c. To protect metals from galvanic corrosion
   d. To remove surface oxides
4. **Which corrosion prevention method involves applying a protective layer on metal surfaces?**
   a. Cathodic protection
   b. Anodic protection
   c. Corrosion inhibitors
   d. Protective coating
5. **What does the Pilling-Bedworth ratio determine in corrosion science?**
   a. Thickness of oxide layer
   b. Rate of metal loss
   c. Surface area exposed to corrosion
   d. Corrosion potential

Answers to Multiple Choice Questions

1. (d) 2. (d) 3. (c) 4. (d) 5. (a)

REFERENCES

1. Amin, M. A., & Ibrahim, M. M. (2011). Corrosion and corrosion control of mild steel in concentrated H_2SO_4 solutions by a newly synthesized glycine derivative. *Corrosion Science, 53*(3), 873–885.

2. Anderson, O. L., & Grew, P. C. (1977). Stress corrosion theory of crack propagation with applications to geophysics. *Reviews of Geophysics, 15*(1), 77–104.
3. Arurault, L. (2008). Pilling–Bedworth ratio of thick anodic aluminium porous films prepared at high voltages in H2SO4 based electrolyte. *Transactions of the IMF, 86*(1), 51–54.
4. Casillas, N., Charlebois, S., Smyrl, W. H., & White, H. S. (1994). Pitting corrosion of titanium. *Journal of the Electrochemical Society, 141*(3), 636.
5. Cavallaro, L., Felloni, L., Trabanelli, G., & Pulidori, F. (1964). The anodic dissolution of iron and the behaviour of some corrosion inhibitors investigated by the potentiodynamic method. *Electrochimica Acta, 9*(5), 485–494.
6. Cowley, W. E., Robinson, F. P. A., & Kerrich, J. E. (1968). Anodic protection for the control of corrosion fatigue. *British Corrosion Journal, 3*(5), 223–237.
7. Cui, Y., Ding, Y., Xu, S., Yang, Z., Zhang, P., Rao, W., & Liu, J. (2018). Liquid metal corrosion effects on conventional metallic alloys exposed to eutectic gallium–indium alloy under various temperature states. *International Journal of Thermophysics, 39*, 1–14.
8. Doig, P., & Flewitt, P. E. J. (1979). A finite difference numerical analysis of galvanic corrosion for semi-infinite linear coplanar electrodes. *Journal of The Electrochemical Society, 126*(12). 2057.
9. Elliott, P. (2003). Materials selection for corrosion control. In *Corrosion: Fundamentals, Testing, and Protection* (Vol. 2, pp. 909–928). ASM International.
10. Ferrando, W. A. (1989). Review of corrosion and corrosion control of magnesium alloys and composites. *Journal of Materials Engineering, 11*(4), 299–313.
11. Glass, G. K., Page, C. L., & Short, N. R. (1991). Factors affecting the corrosion rate of steel in carbonated mortars. *Corrosion Science, 32*(12), 1283–1294.
12. Gong, K., Wu, M., & Liu, G. (2020). Comparative study on corrosion behaviour of rusted X100 steel in dry/wet cycle and immersion environments. *Construction and Building Materials, 235*, 117440.
13. Hœrlé, S., Mazaudier, F., Dillmann, P., & Santarini, G. (2004). Advances in understanding atmospheric corrosion of iron. II. Mechanistic modelling of wet–dry cycles. *Corrosion Science, 46*(6), 1431–1465.
14. Honey, F. J., Kedward, E. C., & Wride, V. (1986). The development of electrodeposits for high-temperature oxidation/corrosion resistance. *Journal of Vacuum Science & Technology A: Vacuum, Surfaces, and Films, 4*(6), 2593–2597.
15. Hu, Q., Zhang, G., Qiu, Y., & Guo, X. (2011). The crevice corrosion behaviour of stainless steel in sodium chloride solution. *Corrosion Science, 53*(12), 4065–4072.
16. Huang, H., Pan, Z., Qiu, Y., & Guo, X. (2013). Electrochemical corrosion behaviour of copper under periodic wet–dry cycle condition. *Microelectronics Reliability, 53*(8), 1149–1158.
17. Jiang, G., Keller, J., & Bond, P. L. (2014). Determining the long-term effects of H2S concentration, relative humidity and air temperature on concrete sewer corrosion. *Water Research, 65*, 157–169.

18. Landolfo, R., Cascini, L., & Portioli, F. (2010). Modeling of metal structure corrosion damage: A state of the art report. *Sustainability, 2*(7), 2163–2175.

19. Loto, C. A. (2017). Microbiological corrosion: Mechanism, control and impact—A review. *The International Journal of Advanced Manufacturing Technology, 92*(9), 4241–4252.

20. Mahesh, B. V., & Raman, R. S. (2014). Role of nanostructure in electrochemical corrosion and high temperature oxidation: A review. *Metallurgical and Materials Transactions A, 45*, 5799–5822.

21. McMahon, M. E., Santucci Jr, R. J., Glover, C. F., Kannan, B., Walsh, Z. R., & Scully, J. R. (2019). A review of modern assessment methods for metal and metal-oxide based primers for substrate corrosion protection. *Frontiers in Materials, 6*, 190.

22. Miyasaka, M., Kishimoto, K., & Aoki, S. (1991). Study on differential aeration cell corrosion owing to differential flow rate analysis by the boundary element method. *Zairyo-to-Kankyo, 40*(6), 401–405.

23. Munn, R. S., & Devereux, O. F. (1991). Numerical modeling and solution of galvanic corrosion systems: part I. Governing differential equation and electrodic boundary conditions. *Corrosion, 47*(8), 612–618.

24. Nam, N. D., Van Hien, P., Hoai, N. T., & Thu, V. T. H. (2018). A study on the mixed corrosion inhibitor with a dominant cathodic inhibitor for mild steel in aqueous chloride solution. *Journal of the Taiwan Institute of Chemical Engineers, 91*, 556–569.

25. Nazari, M. H., Zhang, Y., Mahmoodi, A., Xu, G., Yu, J., Wu, J., & Shi, X. (2022). Nanocomposite organic coatings for corrosion protection of metals: A review of recent advances. *Progress in Organic Coatings, 162*, 106573.

26. Pan, T. J., Al-Omary, M., Zeng, C. L., Wu, W. T., & Niu, Y. (2005). The corrosion of three commercial steels in reducing atmospheres containing HCl and H2S at 873 K. *High Temperature Materials and Processes, 24*(3), 193–203.

27. Park, I. C., & Kim, S. J. (2020). Determination of corrosion protection current density requirement of zinc sacrificial anode for corrosion protection of AA5083-H321 in seawater. *Applied Surface Science, 509*, 145346.

28. Priyantha, N., Jayaweera, P., Sanjurjo, A., Lau, K., Lu, F., & Krist, K. (2003). Corrosion-resistant metallic coatings for applications in highly aggressive environments. *Surface and Coatings Technology, 163*, 31–36.

29. Saurbier, K., Schultze, J. W., & Geke, J. (1994). Temporary inhibitors of corrosion in wet atmosphere: electrochemical investigations of the mechanism and efficiency. *Electrochimica Acta, 39*(8, 9), 1171–1178.

30. Schock, M. R. (1989). Understanding corrosion control strategies for lead. *Journal-American Water Works Association, 81*(7), 88–100.

31. Shao, F., Yang, K., Zhao, H., Liu, C., Wang, L., & Tao, S. (2015). Effects of inorganic sealant and brief heat treatments on corrosion behavior of plasma sprayed Cr_2O_3–Al_2O_3 composite ceramic coatings. *Surface and Coatings Technology, 276*, 8–15.

32. Singh, H., Sidhu, B. S., Puri, D., & Prakash, S. (2007). Use of plasma spray technology for deposition of high temperature oxidation/corrosion resistant coatings–A review. *Materials and Corrosion, 58*(2), 92–102.

33. Soltanloo, M., Babaee, M. H., Yeganeh, S. E. H., Shafiei, A., & Akbari, A. H. (2022). Root cause failure investigation of a boiler waterwall tube employed in a 325 MW thermal power plant: caustic corrosion phenomenon and its effects. *Engineering Failure Analysis, 133*, 105974.

34. Stack, M. M., Abdelrahman, S. M., & Jana, B. D. (2010). A new methodology for modelling erosion–corrosion regimes on real surfaces: Gliding down the galvanic series for a range of metal-corrosion systems. *Wear, 268*(3, 4), 533–542.

35. Wang, Y., Cheng, G., Wu, W., Qiao, Q., Li, Y., & Li, X. (2015). Effect of pH and chloride on the micro-mechanism of pitting corrosion for high strength pipeline steel in aerated NaCl solutions. *Applied Surface Science, 349*, 746–756.

36. Yang, Z., Xia, G. G., Walker, M. S., Wang, C. M., Stevenson, J. W., & Singh, P. (2007). High temperature oxidation/corrosion behavior of metals and alloys under a hydrogen gradient. *International Journal of Hydrogen Energy, 32*(16), 3770–3777.

37. Zhang, Y., Zhang, T., Wan, H., Li, G., & Liu, H. (2023). Mechanism of microbiologically induced anaerobic water-line corrosion of 980 high strength steel in nutrient-rich artificial seawater. *Corrosion Science, 220*, 111268.

38. Zhu, F., Persson, D., Thierry, D., & Taxen, C. (2000). Formation of corrosion products on open and confined zinc surfaces exposed to periodic wet/dry conditions. *Corrosion, 56*(12), 1256–1265.

CHAPTER

8 Green Chemistry

LEARNING OBJECTIVES

After studying this chapter, you will be able to:

1. Understand the 12 principles of Green Chemistry and their importance in sustainable chemical practices.
2. Explain the concepts of waste prevention and atom economy in chemical processes.
3. Discuss methods for synthesizing chemicals in ways that reduce hazardous waste and environmental impact.
4. Identify safe solvent alternatives and their role in promoting green chemistry practices.
5. Describe strategies for designing safer chemicals to minimize human and environmental toxicity.
6. Analyze approaches for enhancing energy efficiency in chemical processes.
7. Evaluate the concept of reducing derivatives in chemical synthesis and its benefits.
8. Explore the role of catalysis in green chemistry, including its applications and benefits.

KEY TERMS FROM THIS CHAPTER

Accident prevention
Biodegradation
Designing safer chemicals
Renewable feedstocks
Waste prevention
Atom economy
Catalysis
Energy efficiency
Safe solvents

INTRODUCTION

The concept of green chemistry is centered on 12 principles aimed at reducing or eliminating hazardous substances throughout the synthesis, production, and use of chemical products. This approach decreases the use of materials that are harmful to human health and the environment. The notion of green chemistry was first articulated in the early 1990s. The 12 principles of green chemistry serve as design principles to help chemists achieve sustainability (Kalidindi and Jagirdar, 2012). Green chemistry is characterized by the careful planning of chemical production and molecular design to minimize adverse effects. Through effective design strategies, synergistic outcomes can be achieved instead of simple trade-offs. Only chemicals and processes that do not harm the environment are employed. The aim is to develop or mimic molecules, reactions, substances, and processes that are initially created or reproduced to be safer for both people and the environment. Green chemistry encompasses a wide range of subfields within the field of chemistry, including organic, inorganic, and biochemical sciences, as well as polymer, ecological, and toxicological studies (Tucker, 2006). It is possible to realize environmental protection and economic benefits through a variety of green program developments, such as catalysis, bio-catalysis, the utilization of sustainable feedstocks, reaction solvents, reaction conditions (e.g., microwave irradiation), and new synthetic pathways (e.g., photocatalytic reactions). This chapter covers recent developments in green chemistry, which lessen the negative influence that chemical procedures and equipment have on the environment (Figure 8.1) (Anastas, 2013). There has been a substantial increase in economic growth and an improvement in living standards as a direct result of the enormous scientific and technological advancements that occurred in the later part of the 20th century. However, economic expansion has also resulted in significant environmental deterioration, as demonstrated by the escalation of climate change, the creation of ozone holes, and the dissemination of organic contaminants that are not biodegradable throughout the biosphere (Anastas and Eghbali, 2010).

Figure 8.1. *Illustration of reducers by green chemistry.*

(Source: Geeks for Geeks, Creative Commons License).

The current situation demands finding a balance between natural resource use, economic growth, and environmental preservation. In the past 20 years, awareness of the importance of environmental protection has grown, resulting in a greater emphasis on green and sustainable technologies. Green chemistry, a burgeoning field in chemistry, focuses on adopting ecological methods to minimize or eradicate the utilization of detrimental compounds in chemical processes and the production of hazardous and toxic byproducts. For a reaction to be classified as "green," it must consist of these essential green elements: a reagent/catalyst solvent, and careful consumption of energy. Furthermore, green chemistry focuses on optimizing waste disposal methods and developing strategies for the degradation of chemical products after they have been used (Clark, 2006). These efforts are aligned to prevent pollution and promote sustainable development. Furthermore, the notion of green chemistry has had a significant influence as it extends beyond the confines of the research laboratory and has made an impact on several sectors such as commerce, environment, schooling, and the general public. The discipline of green chemistry has elucidated the methods by which chemists might create advanced goods and processes that are economically viable while also promoting human well-being and environmental sustainability. In addition to the growing scientific interest in green chemistry, there has been a significant increase in teaching programs, government financing, and the creation of green chemistry research centers over the last 20 years. Currently, several universities provide courses on this subject (Horvath and Anastas, 2007).

Chemistry has traditionally been seen as a hazardous field, with the public often equating the term "chemical" with "toxic." However, risks can be mitigated through the implementation of safety measures, such as using protective equipment. When safety measures fail, the level of risk, which is determined by the combination of the potential danger and the level of exposure, rises. In the event of a high hazard and the failure of exposure measures, the consequences can be catastrophic, potentially resulting in death. To mitigate the risk in situations like accidents, spills, and sabotage, it is important to prioritize minimizing the hazardous elements of the equation rather than only relying on exposure controls (Zimmerman et al., 2020).

The inherent risk associated with a chemical product or process can be systematically minimized at each stage, encompassing factors such as toxicity, physical dangers like explosions and flammability, as well as global risks like stratospheric ozone depletion. The risks associated with these hazards may be heightened due to the characteristics of the feedstock. To remove inherent dangers associated with chemicals and processes, it is possible to achieve this through meticulous design. This design should be based on the incorporation of the 12 principles as a unified and interconnected set (Linthorst, 2010).

8.1. THE TWELVE PRINCIPLES

In 1998, Paul Anastas and John Warner introduced the 12 principles of green chemistry. These principles provide a comprehensive framework for designing new chemical products and processes (Krasnodębski, 2024), covering all stages of the process life cycle. This includes the selection of raw materials, the optimization of efficiency and safety during transformation, the assessment of the harmful effects and biodegradation of products and reagents, and others (Fantoni et al., 2022).

Remember

Remember the Twelve Principles of Green Chemistry, which provide guidelines for reducing environmental impact throughout chemical processes.

8.2. PREVENTION OF WASTE

The primary premise of green chemistry is the prevention of waste, as preventing waste production is more beneficial to health than cleaning it up afterward. Waste can be defined as any material produced that lacks realized value or represents the loss of unused energy. Waste can take various forms and impact the ecosystem differently based on its toxicity, composition, volume, or disposal method (Mulholland et al., 2000).

The rationale for implementing this principle is underscored by the fact that over 12 billion metric tons of waste, including about 300 million metric tons of hazardous waste posing risks to people and the environment, are generated each year in the US (Ivanković et al., 2017). The chemical sector is responsible for generating 70% of the overall quantity of hazardous waste, including a significant volume of organic toxic waste. Specifically, methanol, and xylenes are the predominant substances in this category, accounting for approximately 150,000 tons (Chen et al., 2020). Waste sanitation accounts for approximately 2.2% of the gross US product and its expenditures are consistently increasing. Organic waste, which poses risks to both individuals and the environment, is mainly generated during specific stages of synthesis. It is widely recognized that the use of toxic reactants and solvents in certain reactions leads to the production of a significant amount of toxic substances due to severe reaction conditions. Organic synthesis commonly involves several fundamental reactions such as halogenation, alkylation, oxidation, nitration, and sulfonation (Koel and Kaljurand, 2006). These processes find applications in several industrial sectors. While the chemical industry, along with other chemical makers, has hitherto neglected prevention, the primary focus of green chemistry is to halt the production of waste. Nevertheless, achieving complete waste prevention in reality is almost unattainable due to the inability to fully utilize all input raw materials (Mooney, 2004). On the other hand, the disposal of a single trash exemplifies the ultimate depletion of material resources in the circular process of production and consumption. Hence, the reintegration of material goods into a circular flow signifies a significant economic advantage. It is crucial to consider the feasibility of waste prevention and, if not feasible, to strategize on how to optimize the utilization of waste generated during production, thereby rendering it valuable (Sharma et al., 2008).

8.3. ATOM ECONOMY

In 1990, Barry Trost established the notion of synthetic efficiency, specifically atom economy (AE), commonly referred to as atom efficiency (Murphy, 2023). This concept focuses on maximizing the utilization of raw materials. For example, the result comprises the highest possible quantity of atoms derived from the initial substances. A perfect reaction would involve the complete incorporation of all the atoms from the reactants (Wang et al., 2011). The AE, or Atom Economy, is determined by dividing the molecular weight of the desired product by the combined molecular weights of all the reactants involved in the reaction (*see* Figure 8.2). The term "theoretical value" refers to a value that is used to quickly evaluate the efficiency of a reaction (Cann and Dickneider, 2004).

$$AE = \frac{\text{MW of Product}}{\text{MW of Reagents}}$$

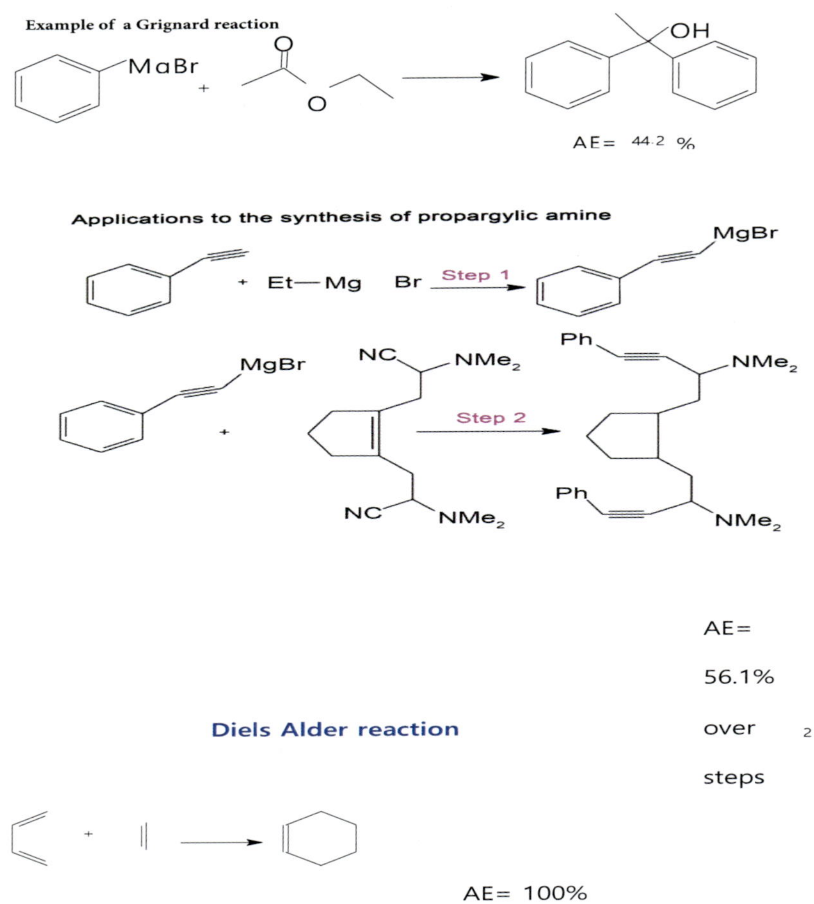

Figure 8.2. *The atom economy AE and several demonstrations [3].*

8.4. CHEMICAL SYNTHESIS FOR LESS HAZARDOUS

Several novel reactions have been created over the past decade, supplementing the preexisting green reactions. Reactions such as rearrangement cycloaddition or multi-component reaction of coupling have already been discovered and fall within the category of efficient processes (Gangrade et al., 2015). Numerous industrial processes have become cleaner and more cost-effective by substituting hazardous chemicals with natural enzymes. For example, a new polycarbonate synthesis (PC) method has been developed that is fundamentally straightforward, replacing hazardous $COCl_2$ with CO_2. This process also results in the elimination of CH_2Cl_2 as a solvent. The overall reaction involves the combination of ethylene oxide $(CH_2)_2O$, CO_2, and bisphenol-A ($C1_5H_{16}O_2$) to form $C_2H_6O_2$, as depicted in Figure 8.3 (Grewal et al., 2013).

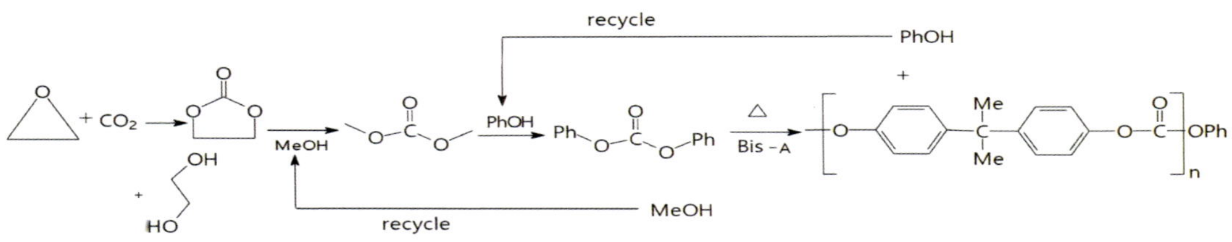

Figure.8.3. *Synthesis of polycarbamate [25].*

8.5. SAFE SOLVENTS

Whenever feasible, the utilization of compounds such as solvents and separating agents should be minimized and rendered harmless. Chromatographic separations, for instance, necessitate significant amounts of solvents, which pose significant challenges as sources of environmental pollution. The majority of conventional organic solvents exhibit toxicity, flammability, and corrosiveness. Their recycling process is linked to energy-efficient distillation, however, there are significant losses involved (Gu and Jérôme, 2010). Hence, there is a need for the advancement of eco-friendly solvents. Safer Solvents advises minimizing the synthesis process and, whenever feasible, avoiding the usage of auxiliary chemical substances. If such substances are necessary, they should be non-toxic. The selection of appropriate replacements for organic solvents in accordance with the principles of green chemistry is determined by considerations of process safety, employee safety, environmental safety, and the long-term viability of the process. The solvent should possess physical and chemical stability, exhibit low volatility, be user-friendly, and be readily recyclable (Dutta et al., 2022).

The field of green research in chemistry is mostly focused on solvents. An essential difficulty for green chemistry is the significant amount of mass lost in synthesis and processing, which is often dominated by these substances. In addition, numerous traditional solvents possess toxicity, flammability, and corrosiveness. Their volatility and solubility have resulted in the contamination of air, water, and land, hence elevating the likelihood of workers being exposed and causing severe accidents (Welton, 2015). Whenever feasible, the process of recovering and reusing materials is frequently associated with energy-intensive distilling and occasionally leads to cross-contamination. In order to rectify these deficiencies, scientists embarked on a quest to find alternative solutions that are less hazardous (Nanda et al., 2021).

Did you know?

Did you know that safe solvents in green chemistry are designed to minimize toxicity and environmental impact while maintaining effective chemical reactions?

Ideally, it would be preferable to avoid using any solvent whenever feasible, as the inclusion of an auxiliary always necessitates exerting efforts and energy to eliminate it from a chosen system. Consequently, efforts have been dedicated to developing systems that do not require solvents (Claux et al., 2021).

H_2O, the most prevalent molecule on Earth, is often regarded as a versatile and harmless solvent. Conducting reactions in water offers significant benefits. Water is a benign, non-hazardous substance that can serve as a useful solvent in large-scale chemical reactions. The unique properties of water, such as the hydrophobic effect and its ability to prevent the dissolution of many organic molecules, have led to enhanced reaction rates and improved separation processes. An example that illustrates the benefits of using water as a solvent is the improved Diels-Alder process conducted in water. However, an unresolved drawback that can hinder industrial applications is the potential for water pollution, which necessitates substantial energy expenditure for remediation (Cvjetko Bubalo et al., 2015).

8.6. DESIGNING SAFER CHEMICALS

Chemicals should be designed to optimize their desired functionality while minimizing their toxicity. Creating safer products and processes involves the challenging task of reducing toxicity without sacrificing function and efficiency (Sheldon, 2012). This requires a comprehensive understanding of chemistry, toxicology, and ecological science. The goal of manufacturing chemicals that do not cause cancer, genetic mutations, or nervous system damage is achieved by balancing maximum efficiency with the intended function of the chemical product. This ensures the toxicity and potential danger are minimized as much as possible (Karagölge and Gür, 2016).

This approach has been utilized in the development of new insecticides and pesticides that specifically target certain organisms. These chemicals are toxic only to the intended targets and break down into environmentally benign compounds. For instance, extremely poisonous organic tin-based compounds (Sn), previously used as coatings to prevent seaweed and plankton from attaching to surfaces, have been replaced with a non-toxic substance known as Sea-Nine. Hydrogen peroxide can serve as an alternative to chlorine bleaches in paper manufacturing, helping to avoid the use of ozone-depleting chemicals (Beach et al., 2009).

Another instance involves the manufacturing of polyphenylene sulfone (PPSU) polymers, which are currently extensively utilized in indoor aircraft and have also been used in underground trains as a crucial measure to employ non-combustible materials. This is a recently developed engineering plastic that possesses a unique combination of beneficial ecological, mechanical, and flame-resistant qualities (Morales et al., 2020).

8.7. DESIGN FOR ENERGY EFFICIENCY

Rising apprehensions about the exhaustion of petroleum feedstocks and the rise in energy use have spurred the advancement of more energy-efficient methods and the quest for renewable energies; resources that do not deplete inside a timeframe that is significant to human existence. It is important to acknowledge the energy demands of chemical processes due to their significant economic and environmental consequences (Ruiz-Mercado et al., 2016). The oil crisis in 1973 prompted the creation of several procedures that prioritize energy conservation and aim to maximize the utilization of every kilojoule of energy during the production process. Lowering the activation energy of a chemical process or selecting appropriate reactants to enable the reaction to occur at ambient temperature is an instance of how chemists can minimize the energy demands (Sáenz-Galindo et al., 2020), resulting in various direct and indirect advantages. Increasing the energy effectiveness of a chemical system is only one component of the answer (Sheldon, 2016). There is a demand for alternative forms of energy. Various forms of renewable energy have been acknowledged for their contribution to biofuel production, such as solar power, windmills, geothermal power, hydropower, and fuel cell technology for hydrogen (Abdussalam-Mohammed et al., 2020).

Solar energy is often regarded as the foremost renewable energy source on the planet and is seen as a viable substitute for petroleum. The fundamental concept behind these cells is the material's ability to capture and convert photonic energy from sunlight. When light is absorbed, it creates excited states that can transfer energy and generate an electric current. The efficient conversion of light into electrical current using building materials remains a difficulty and is crucial for the success of this technique (Slimani and Hannachi, 2021).

8.8. REDUCE DERIVATIVES

It is advisable to minimize or eliminate unnecessary derivatization techniques, such as the use of blocking groups, temporary alteration of physical/chemical processes, and protection/deprotection. These steps need additional reagents and result in the production of waste. If feasible, the process of biological synthesis must be employed.

For instance, a non-covalent protective group was created by forming a co-crystal comprising hydroquinones and bis(N, N-dialkyl)terephthalamides (Figure 8.4). This method proved to be effective and successful for industrial applications (Molnar et al., 2020).

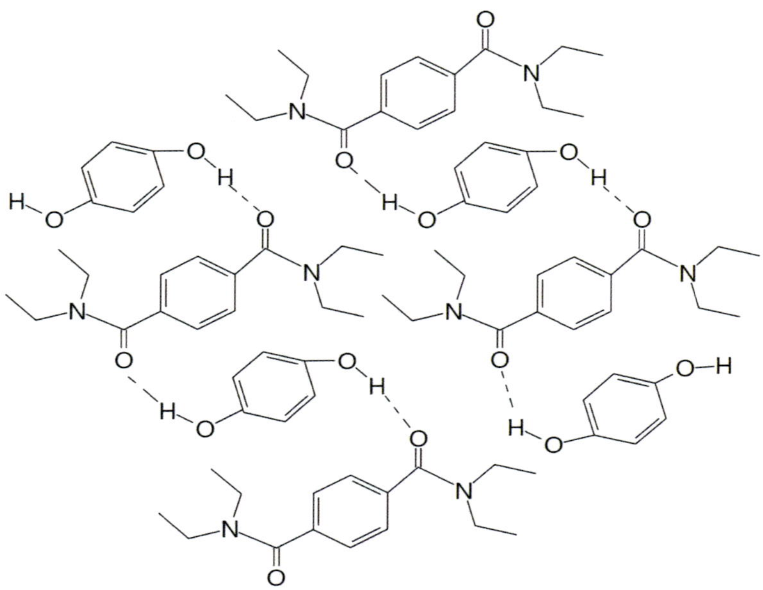

Figure 8.4. *Illustrates 8 hydroquinones protected by non-covalent interactions with bis-(N, N-dialkyl)terephthalamides [3].*

8.9. CATALYSIS

When compared to stoichiometric reagents, catalytic reagents offer a greater number of benefits. The formation of waste is frequently linked to the customary usage of a stoichiometric quantity of chemicals. This is a fact that is widely acknowledged by the scientific community.

For the purpose of protecting the environment, the idea of catalysis encourages the employment of catalysts that are biodegradable. This strategy results in a decrease in the amount of energy that is consumed, results in the elimination of the requirement for organochlorine compounds, and reduces the amount of wastewater and energy waste that is produced (Min and Friend, 2007).

In addition, biocatalysis is an example of green chemistry because it employs a biomimetic technique that makes use of enzymes that are either naturally occurring or engineered. In most cases, it denotes the consumption of enzymes that are in their purest form, in addition to the alterations that are carried out by organisms that have been genetically modified. As an additional point of interest, the circumstances of the reaction are quite mild, as the conversion can be carried out in water at room temperature and under normal air pressure. A further point to consider is that enzymes have been shown to have enhanced stereochemistry and chemical selectivity. Furthermore, in comparison to non-biological catalysts, biocatalysts offer a number of major benefits, including increased reaction rates, decreased costs, and increased catalytic specificity. On the other hand, they are restricted due to the fact that they are not sensitive to heat and exhibit unstable behavior (Liu et al., 2015).

Tip

Catalysis plays a crucial role in green chemistry by enhancing reaction rates and reducing the need for harsh reaction conditions and hazardous chemicals.

8.10. BIODEGRADATION

The formulation of chemical products ought to be such that they degrade into harmless degradation products without causing any harm to the environment. When they have finished performing their intended purpose, the goal is to create compounds and products that are biodegradable and capable of transforming into substances that are safe for the environment upon completion. The implementation of recycling is intended to achieve the goal of minimizing the production of harmful substances and maximizing the amount of waste that may be reused.

The problem of endurance has been recognized for a substantial amount of time, and it became apparent during the early stages of the development of the industrial sector. Tetra propylene-alkylbenzene sulfonate, often known as TPPS, was utilized as a surfactant in laundry detergents that were manufactured in the early 1950s. Despite this, TPPS would accumulate in the water supply since the degrading process was not completed. Because the water had a tendency to foam up as soon as it was released from the tap, the situation was quite dangerous.

As seen by the ongoing problems of environmental contamination, the process of developing compounds and chemicals that are capable of degrading by themselves is a difficult task. After a number of years of data collection, patterns have become abundantly clear. It is best to steer clear of chemical compounds that have branched chains, quaternary carbons, halogenated moieties, tertiary amines, and particular heterocycles because they have the potential to demonstrate enhanced persistence. Alternately, the incorporation of functional groups that are recognized by common enzymes, such as esters or amides, could be of assistance in the synthesis of molecules that are environmentally degradable (Jordan et al., 2016).

8.11. ANALYSIS

Green analytical chemistry is the application of analytical methods that produce less waste and provide minimal risks to both the environment and people. This description takes into account both the continuing monitoring of a chemical change as well as the environmental problems that are linked with conventional analysis.

The purpose of green analytical chemistry is to ascertain the quantity of chemicals while simultaneously reducing or doing away with the production of waste. There is a primary connection between the analytical technique itself and the environmental concerns that are related to analytical chemistry. It is important to take into consideration the items that are used in the manufacturing of analytical instruments. Green engineers and chemists who are working on the development of novel sensors should be aware of the toxicity of the chemical they are manipulating as well as the potential environmental difficulties that are associated with it. Electrode materials made of mercury are utilized quite commonly in the field of electrochemistry today (Melchert et al., 2012).

8.12. ACCIDENT PREVENTION

A chemical process should give priority to avoiding the likelihood of chemical accidents, such as leaks, explosions, and fires while selecting the substances that will be used in the process involving chemicals. For example, the increasing use of supercritical carbon dioxide (CO_2) as an alternative to organic solvents, which are neither dangerous nor explosive and have a positive effect on the environment, is a good example. Managing known risks in order to achieve a level of risk that is acceptable is the act that constitutes safety.

The use of personal protective equipment is reduced to the greatest extent possible in order to achieve this goal. The identification of potential hazards is the initial stage in accident prevention, as stipulated by the modifications to the Chemical Accident Prevention and Clean Air Act that were passed in 1990. When designing chemicals and processes, potential hazards must be considered, such as toxicity, physical risks like explosivity or flammability, and global hazards. This is absolutely necessary in order to forestall calamities that are comparable to the Love Canal tragedy and the Bhopal tragedy. Considering that numerous chemicals continue to pose major hazards and ought to be replaced with safer alternatives in order to proactively prevent disasters wherever it is feasible, this tragedy serves as an important wake-up call for the scientific community (Schulte et al., 2013).

8.13. USE OF RENEWABLE FEEDSTOCKS

In the field of green chemistry, one of the fundamental principles advocates for the usage of renewable feedstocks wherever it is feasible from both an economic and a technological standpoint. As an illustration, it is recommended to use renewable raw resources as an alternative to a variety of plastic substances, and then to dispose of the waste materials after completion of the process. Because of this particular reason, the manufacture of biodegradable plastics is currently a trend that is gaining popularity. There is a bright future ahead for the food industry in terms of the adoption of biodegradable packaging. There are a number of elements that have an impact on the development of biodegradable packaging, including politics, changes in regulation, and the global need for both food and energy resources. Over the years, there have been major breakthroughs made in the manufacture of fuels, chemicals, and goods that get their materials from renewable resources. When it comes to bioplastics, for instance, the utilization of sustainable raw materials throughout the production process has a positive impact on the amount of energy that is consumed as well as the amount of carbon dioxide that is released. At the moment, the multinational corporation known as Coca-Cola manufactures bottles that are made up of a mixture that contains 30 % polyethylene (PE) (Gallezot, 2007).

8.14. DISADVANTAGES OF GREEN CHEMISTRY

In the field of green chemistry, the primary purpose is to produce chemical products and procedures that reduce or eliminate the use of harmful chemicals. In addition, this purpose is the primary emphasis of green chemistry, which is obvious in the impact that it has on time, expenses, and the lack of information. When switching from a traditional product to an ecologically friendly one, the process of designing and developing the new product as well as the manufacturing process for the new product can be difficult and expensive. In addition, there is a lack of consensus over what constitutes safe behavior, and there may be no recognized replacements for some chemicals or raw materials (Ziolek, 2004).

Because of the high cost of implementation and the limited amount of knowledge available, there is a possibility that green chemical techniques will not be adopted to the extent that they should be. The consequence of this could be that there are not enough alternatives that have been developed for the use of chemical raw materials or alternative technologies in environmentally friendly operations. In addition, there is a severe lack of capabilities among human beings. In the realm of ecologically friendly chemistry, ionic liquids are representative of the future. Without a doubt, such are helpful in the process of chemical synthesis. When the 12 principles that make up green chemicals are taken into consideration, it is possible to make the observation that ionic liquids do not appear to be ecologically friendly by their nature. Despite the fact that it is common knowledge that ionic liquids have a low vapor pressure, which leads to a low level of volatility, this is only one of the many variables that contribute to a substance being deemed environmentally friendly. It is anticipated that liquids based on ions, fluoro-anion, and imidazole will be hazardous; nevertheless, these liquids are unable to reach the environment through the process of evaporation from the environment. Because the vast majority of ionic liquids are capable of dissolving in water, they are able to quickly enter the biosphere through this pathway (Pal et al., 2022).

ACTIVITY 8.1.

Use of Renewable Feedstocks): Activity: Explore the use of renewable feedstocks in chemical synthesis by researching and comparing traditional and green chemistry approaches to production processes.

SUMMARY

The objective of green chemistry, as previously stated, is to adhere to 12 principles that seek to minimize the use of hazardous substances in the manufacturing and use of chemical goods. It is widely recognized that it is not feasible to simultaneously fulfill all 12 principles of a green chemical process when constructing it. However, it strives to incorporate as many concepts as feasible throughout specific stages of synthesis. The objective of green chemistry, which aims to protect the environment and generate economic benefits, can be accomplished through several approaches. For example, chemical products would be formulated in such a way that they do not linger in the environment and instead undergo complete breakdown into harmless byproducts.

The primary objective of green chemistry is to fulfill societal needs while minimizing the use of damaging or decreasing natural resources on Earth. In this instance, the emphasis is being redirected towards the production of items that can be completely recovered or reused. Green chemistry aims to achieve the key goal of reducing pollution and waste by altering production and consumption habits. It is imperative to cultivate alternate technologies in order to avert any additional harm to both human health and the environment. Environmental conservation can be achieved by implementing a green chemistry method. While green chemistry is important, it is not sufficient to address the urgent environmental issues and impacts of the current period. However, implementing the 12 principles of green chemistry in practical applications will ultimately contribute to creating a better world.

REVIEW QUESTIONS

1. What are the 12 Principles of Green Chemistry, and why are they important for sustainable chemical practices?
2. Describe the concept of atom economy and its significance in reducing waste generation in chemical reactions.
3. Discuss the advantages and challenges of using renewable feedstocks in chemical synthesis.
4. Explain the role of catalysis in green chemistry and provide examples of catalytic processes that improve efficiency and sustainability.
5. How does green chemistry promote the use of safer solvents? Provide examples of safer solvent alternatives.

6. Analyze the role of biodegradation in green chemistry and its implications for environmental sustainability.
7. What are the potential disadvantages or barriers to implementing green chemistry practices on a broader scale? How can these be addressed?

MULTIPLE CHOICE QUESTIONS

1. Which principle of green chemistry focuses on minimizing waste generation?
- a. Atom economy
- b. Design for energy efficiency
- c. Prevention
- d. Catalysis

2. What is the primary goal of safe solvents in green chemistry?
- a. Enhance chemical reactivity
- b. Reduce waste
- c. Improve safety in chemical processes
- d. Increase product yield

3. Which principle advocates for designing chemical processes that use renewable feedstocks?
- a. Biodegradation
- b. Prevention
- c. Use of renewable feedstocks
- d. Analysis

4. What is the main disadvantage associated with green chemistry?
- a. Limited application in pharmaceuticals
- b. Complexity in design
- c. Higher costs in chemical synthesis
- d. Potential reduction in chemical efficiency

5. What is the purpose of catalysis in green chemistry?
- a. To increase reaction temperature
- b. To enhance product yield
- c. To speed up chemical reactions
- d. To reduce environmental impact

Answers to Multiple Questions

1. (a) 2. (c) 3. (c) 4. (d) 5. (c)

REFERENCES

1. Abdussalam-Mohammed, W., Ali, A. Q., & Errayes, A. O. (2020). Green chemistry: Principles, applications, and disadvantages. *Chemistry Methodology, 4*(4), 408–423.
2. Anastas, P. T. (2013). *Handbook of Green Chemistry* (Vol. 1, pp. 21–25). Wiley-VCH.
3. Anastas, P., & Eghbali, N. (2010). Green chemistry: Principles and practice. *Chemical Society Reviews, 39*(1), 301–312.
4. Beach, E. S., Cui, Z., & Anastas, P. T. (2009). Green chemistry: A design framework for sustainability. *Energy & Environmental Science, 2*(10), 1038–1049.
5. Cann, M. C., & Dickneider, T. A. (2004). Infusing the chemistry curriculum with green chemistry using real-world examples, web modules, and atom economy in organic chemistry courses. *Journal of Chemical Education, 81*(7), 977.
6. Chen, T. L., Kim, H., Pan, S. Y., Tseng, P. C., Lin, Y. P., & Chiang, P. C. (2020). Implementation of green chemistry principles in circular economy system towards sustainable development goals: Challenges and perspectives. *Science of the Total Environment, 716*, 136998.
7. Clark, J. H. (2006). Green chemistry: Today (and tomorrow). *Green Chemistry, 8*(1), 17–21.
8. Claux, O., Santerre, C., Abert-Vian, M., Touboul, D., Vallet, N., & Chemat, F. (2021). Alternative and sustainable solvents for green analytical chemistry. *Current Opinion in Green and Sustainable Chemistry, 31*, 100510.
9. Cvjetko Bubalo, M., Vidović, S., Radojčić Redovniković, I., & Jokić, S. (2015). Green solvents for green technologies. *Journal of Chemical Technology & Biotechnology, 90*(9), 1631–1639.
10. Dutta, P., McGranaghan, A., Keller, I., Patil, Y., Mulholland, N., Murudi, V., & Godineau, E. (2022). A case study in green chemistry: The reduction of hazardous solvents in an industrial R&D environment. *Green Chemistry, 24*(10), 3943–3956.
11. Fantoni, T., Tolomelli, A., & Cabri, W. (2022). A translation of the twelve principles of green chemistry to guide the development of cross-coupling reactions. *Catalysis Today, 397*, 265–271.
12. Gallezot, P. (2007). Process options for converting renewable feedstocks to bioproducts. *Green Chemistry, 9*(4), 295–302.
13. Gangrade, D., Lad, S. D., & Mehta, A. L. (2015). Overview on microwave synthesis—Important tool for green chemistry. *International Journal of Research in Pharmacy & Science, 5*(2), 65–68.
14. Grewal, A. S., Kumar, K., Redhu, S., & Bhardwaj, S. (2013). Microwave-assisted synthesis: A green chemistry approach. *International Research Journal of Pharmaceutical and Applied Sciences, 3*(5), 278–285.
15. Gu, Y., & Jérôme, F. (2010). Glycerol as a sustainable solvent for green chemistry. *Green Chemistry, 12*(7), 1127–1138.
16. Horvath, I. T., & Anastas, P. T. (2007). Innovations and green chemistry. *Chemical Reviews, 107*(6), 2169–2173.

17. Ivanković, A., Dronjić, A., Bevanda, A. M., & Talić, S. (2017). Review of 12 principles of green chemistry in practice. *International Journal of Sustainable and Green Energy, 6*(3), 39–48.

18. Jordan, A., Haiß, A., Spulak, M., Karpichev, Y., Kümmerer, K., & Gathergood, N. (2016). Synthesis of a series of amino acid-derived ionic liquids and tertiary amines: Green chemistry metrics including microbial toxicity and preliminary biodegradation data analysis. *Green Chemistry, 18*(16), 4374–4392.

19. Kalidindi, S. B., & Jagirdar, B. R. (2012). Nanocatalysis and prospects of green chemistry. *ChemSusChem, 5*(1), 65–75.

20. Karagölge, Z., & Gür, B. (2016). Sustainable chemistry: Green chemistry. *Journal of the Institute of Science and Technology, 6*(2), 89–96.

21. Koel, M., & Kaljurand, M. (2006). Application of the principles of green chemistry in analytical chemistry. *Pure and Applied Chemistry, 78*(11), 1993–2002.

22. Krasnodębski, M. (2024). The bumpy road to sustainability: Reassessing the history of the twelve principles of green chemistry. *Studies in History and Philosophy of Science, 103*, 85–94.

23. Linthorst, J. A. (2010). An overview: Origins and development of green chemistry. *Foundations of Chemistry, 12*, 55–68.

24. Liu, Y., Zhao, G., Wang, D., & Li, Y. (2015). Heterogeneous catalysis for green chemistry based on nanocrystals. *National Science Review, 2*(2), 150–166.

25. Melchert, W. R., Reis, B. F., & Rocha, F. R. (2012). Green chemistry and the evolution of flow analysis: A review. *Analytica Chimica Acta, 714*, 8–19.

26. Min, B. K., & Friend, C. M. (2007). Heterogeneous gold-based catalysis for green chemistry: Low-temperature CO oxidation and propene oxidation. *Chemical Reviews, 107*(6), 2709–2724.

27. Molnar, M., Lončarić, M., & Kovač, M. (2020). Green chemistry approaches to the synthesis of coumarin derivatives. *Current Organic Chemistry, 24*(1), 4–43.

28. Mooney, D. (2004). Effectively minimizing hazardous waste in academia: The green chemistry approach. *Chemical Health & Safety, 11*(3), 24–28.

29. Morales, C. G., Castolo, A. A., & Rodriguez, M. A. (2020). Designing safer chemicals. In *Green Chemistry and Applications* (Vol. 2, pp. 60–83). CRC Press.

30. Mulholland, K. L., Sylvester, R. W., & Dyer, J. A. (2000). Sustainability: Waste minimization, green chemistry, and inherently safer processing. *Environmental Progress, 19*(4), 260–268.

31. Murphy, M. A. (2023). Professors Trost and Sheldon's promotion of catalytic technologies, atom economy, and the E-factor metrics in synthetic organic chemistry and the fine chemical and pharmaceutical industries, to speed the early evolution of "green chemistry." *Substantia, 7*(2), 41–55.

32. Nanda, B., Sailaja, M., Mohapatra, P., Pradhan, R. K., & Nanda, B. B. (2021). Green solvents: A suitable alternative for sustainable chemistry. *Materials Today: Proceedings, 47*, 1234–1240.

33. Pal, K., Chakroborty, S., & Nath, N. (2022). Limitations of nanomaterials insights in green chemistry sustainable route: Review on novel applications. *Green Processing and Synthesis, 11*(1), 951–964.

34. Ruiz-Mercado, G. J., Carvalho, A., & Cabezas, H. (2016). Using green chemistry and engineering principles to design, assess, and retrofit chemical processes for sustainability. *ACS Sustainable Chemistry & Engineering, 4*(11), 6208–6221.

35. Sáenz-Galindo, A., Cedillo-Portilo, J. J., Espinoza-Cavazos, K. G., de Léon-Martínez, P. A., & Castañeda-Facio, A. O. (2020). Design for energy efficiency. In *Green Chemistry and Applications* (Vol. 2, pp. 95–102). CRC Press.

36. Schulte, P. A., McKernan, L. T., Heidel, D. S., Okun, A. H., Dotson, G. S., Lentz, T. J., & Branche, C. M. (2013). Occupational safety and health, green chemistry, and sustainability: A review of areas of convergence. *Environmental Health, 12*, 1–9.

37. Sharma, S. K., Chaudhary, A., & Singh, R. V. (2008). Gray chemistry versus green chemistry: Challenges and opportunities. *Rasayan Journal of Chemistry, 1*, 68–92.

38. Sheldon, R. A. (2012). Fundamentals of green chemistry: Efficiency in reaction design. *Chemical Society Reviews, 41*(4), 1437–1451.

39. Sheldon, R. A. (2016). Green chemistry and resource efficiency: Towards a green economy. *Green Chemistry, 18*(11), 3180–3183.

40. Slimani, Y., & Hannachi, E. (2021). Green chemistry and sustainable nanotechnological developments: Principles, designs, applications, and efficiency. In *Green Polymer Chemistry and Composites* (Vol. 2, pp. 1–18). Apple Academic Press.

41. Tucker, J. L. (2006). Green chemistry, a pharmaceutical perspective. *Organic Process Research & Development, 10*(2), 315–319.

42. Wang, W., Lü, J., Zhang, L., & Li, Z. (2011). Real atom economy and its application for evaluation of the green degree of a process. *Frontiers of Chemical Science and Engineering, 5*, 349–354.

43. Welton, T. (2015). Solvents and sustainable chemistry. *Proceedings of the Royal Society A: Mathematical, Physical and Engineering Sciences, 471*(2183), 20150502.

44. Zimmerman, J. B., Anastas, P. T., Erythropel, H. C., & Leitner, W. (2020). Designing for a green chemistry future. *Science, 367*(6476), 397–400.

45. Ziolek, M. (2004). Catalytic liquid-phase oxidation in heterogeneous systems as a green chemistry goal: Advantages and disadvantages of MCM-41 used as a catalyst. *Catalysis Today, 90*(1, 2), 145–150.

Index

Kolibri 360 GmbH
Ferdinand-Jühlke-Straße 7
99095 Erfurt, Deutschland
produktsicherheit@kolibri360.de